Sten Hellström

ESD - The Scourge of Electronics

Springer

Berlin
Heidelberg
New York
Barcelona
Budapest
Hong Kong
London
Milan
Paris
Santa Clara
Singapore
Tokyo

Sten Hellström

ESD - The Scourge of Electronics

With 174 Figures and 7 Tables

Springer

Dr. Sten Hellström
RELECTRONIC-ESD-HB
Laxvägen 46
18130 Lidingö
Sweden

ISBN 3-540-61669-1 Springer Verlag Berlin Heidelberg New York

Cataloging-in-Publication Data applied for

Die Deutsche Bibliothek - CIP-Einheitsaufnahme

Hellström, Sten:
ESD - The scourge of electronics / Sten Hellström.- Berlin ; Heidelberg ; New York ; Barcelona ; Budapest ; Hongkong ; London ; Mailand ; Paris ; Santa Clara ; Singapur ; Tokio : Springer, 1998
ISBN 3-540-61669-1

Printed in Germany

Coverdesign: Struve & Partner, Heidelberg
Typesetting: Camera-Ready by author

SPIN: 10548369 62/3020 - 5 4 3 2 1 0 - Printed on acid-free paper

List of contents

Preface

This book on electrostatic discharge phenomena is essentially a translation and update of a Swedish edition from 1992.

The book is intended for people working with electronic circuits and equipments, in application and development. All personnel should be aware of the ESD-hazards, especially those responsible for quality. ESD-prevention is a part of TQM (Total Quality Management). The book is also usable for courses on the subject.

Background

It was soon realised that the MOS-circuits (MOS=Metal Oxide Semiconductor), which appeared in the beginning of the 1960-ties were sensitive to electrostatic discharges. But a severe accident accelerated the search for materials that do not generate electric charges. In April 1964 three people were working inside a satellite at Cape Kennedy Space Center. They suddenly screamed "we are burning". They died. The satellite incapsulation was covered with untreated plastics to protect against dust. When the plastics was pulled off both this and the metal incapsulating got charged. A discharge from the metal ignited inflammable parts of the satellite. Eleven more people were injured and the cost of the accident amounted to about 55 billions USD.

A chemist, Dan Andersson, then at Allied Chemical, later at Richmond Technology, was charged by the defence authorities to develop a plastic, that does not generate electric charges. His method was to use a basic material, polyethene, incorporated with a conducting fluid, that migrates to the surface keeping this somewhat conducting independently of the surrounding humidity. That was the birth of what now is called antistatic materials. To separate this from ordinary plastics, it is given a pink colour and is also known on the market as "pink-poly".

Plastics loaded with carbon (soot) to make it conducting had already been prepared and came now to an extended use.

Another incident illustrates the danger of electrostatic discharge in daily life. A telephone operator got an inquiry, for which she needed her computer to answer. The screen was dusty and when talking in her phone she wiped it with a wet serviette. Suddenly she heard a strong crack in one ear and the phone connection was broken. She had got charged by the wiping. Analysis and simulation of the situation proved, that a discharge took place from an ear-ring through the electronic circuit of the phone to earth. She felt pain, got chocked and her hearing-ability was lowered temporarily!

Electronics and electrical discharges generated by plastics.

In 1968 an American firm manufactured film resistors of high precision to be exported to the German Space industry. The export was executed by an other firm in USA, that made the final inspections. The components were packed in closed cardboards. No failing parts had been discovered at the inspection. But for the components to be handled in clean-rooms they got repacked in clean polyethene bags! Arrival inspection in Germany revealed that out of 5000 items 190 resistance values were changed with as much as 6%, acceptable tolerance being 0.1%!

The wife of the chief engineer in the manufacturing company heard about this and exclaimed: "Surely it is due to that plastics". She was no scientist but had experienced that plastic bags and nylon clothes could cause violent discharges. The chief engineer then made experiments with the plastic bags and the components, rubbing the latter with a bag. Checking the resistance values afterwards showed change of them. Then the long struggle for protective materials started. Pink-poly got new applications as well as carbon loaded plastics, called "carbon black" in this context. Many new materials have been developed see, chapter 11.

Acknowledgements

I am much obliged to my coauthors of the Swedish edition, Mr Per Eklöf of Celsius-Tech, Sweden, and Mr Anders Welander of the Defence Materials Administration, Sweden. Likewise prof. Viktor Scuka, High Voltage Laboratory at the university of Uppsala, Sweden, is recognised for his comments as well as prof. Niels Jonassen, the Technical university of Denmark for discussion on ionization problems. Many worthful experiences were gained from members of the US ESD-Ass. and Swedish colleagues. Dr C. Duvvury and Mr L. Caliciotti of Texas Instruments (Dallas, USA and Avezzano, Italy) gave information on memory submicron technology. Mr M. Honda, Impulse Physics lab., Inc. Japan, is recognised for supplying valuable data on indirect ESD. I thank dr L.O. Eriksson, Institute of Applied Mathematics, Sweden and prof. E. Velmre, dr B. Freydin, dr A. Udal, Technical University of Tallinn, Estonia for collaboration on ESD-simulations. A special thank is given to prof dr A. Birolini of ETH, Zürich, Switzerland for supporting this edition.

Finally I will thank Springer Verlag, Heidelberg and Berlin, Germany, for the publication.

SDR Scandinavian Data Resource AB, Täby, Sweden, made the PC performance of the book.

1 The discovery of static electricity and its manifestation

Already in the stone age our ancestors were fascinated by the mysterious phenomenon of static electricity. Thales from Miletos (mathematician) described about 600 years B.C. how the material amber could become "animated" and attract dust and down. In 1600, William Gilbert, English physician and physicist, published his famous work "De magnete" the first scientific treatment of magnetic and electric phenomena, pointing out their different character. He also postulated that earth is a great magnet.

After the Greek word for amber, "electron", he named materials with similar properties "electrics". Sapphire, diamond, sulphur, resin and glass were of that kind accordingly. Materials that were not able to electrify, metals for instance, he called "non-electrics".

Now it is well known that even metal pieces can keep static electricity when un-earthed. Gilbert invented the electroscope, see Fig. 6, by which static electricity can be indicated.

The first electric machine, a rotating sulphur globule, was constructed about the year 1600 by Otto von Guericke, a German physicist, engineer and diplomate. He is most well known as inventor of the vacuum pump. A famous application of this was the experiments with the Magdeburg hemispheres. When evacuated, eight horses were needed to separate them!

Francis Hauksbee (born 1666), an English engineer and instrument-maker, contributed much to the understanding of static electricity. He was a good experimentalist and is said to be the real discoverer of the frictional electricity (later named "triboelectricity"). He exchanged the sulphur globe of Guericke with a globe or cylinder of glass and intentionally used it for studying electricity. His first paper was however on "barometric light", a phenomenon discovered in 1675. The light is produced when a barometer tube is shaken, intermittent flashes occur in the vacuum above the mercury column. It is due to static electricity created from friction between the glass and the mercury.

Another Englishman, Stephen Gray, discovered in 1729 the difference between conductors and insulators. He also found that the amount of electricity on objects of equal size was independent of their masses. He stated the phenomenon of electric conduction.

The work of Gray was complemented by Charles Dufay, French officer but also interested in natural science. Dufay contributed to electrostatics. He showed in 1734 that there are two types of electricity, i.e. charges, "vitreous" and "resinuous", those of same kind repelling and those of different kind attracting each others. He had correspondence with Gray and also continued Gray's work on conductors, showing that the conductivity varies with materials. He became a member of the French Academy of Science.

The capacitor effect for electric charge was communicated at about the same time by Petrus van Musschenbroek in Holland and Ewald von Kleist in Pommern. The former used together with his coworker Cuneaus in Leiden a glass jar. After pouring water into the jar it received charge from an electrical machine by means of a metal bar. The charged jar discharged when the experimenter touched a metal connector leading to its interior. This invention became known as the Leyden jar. If the jar was instead filled with gold flaws they could increase the effect.

Benjamin Franklin, well known for proving that lightning is an electrical phenomenon, was born 1706 in Boston, Massachusetts, at that time a part of the English colonies in North America. He came to play an important role for the independence of North America. His invention of lightning conductor was based on the observation that metal rods with a peak increased the attraction of lightning. This construction is said to be the first practical application to electricity. He also proposed the one-fluid theory of electricity – an advance to other existing theories and in agreement with most of his experimental results. Franklin's ideas about lightning were confirmed through experiments performed by his French translator, Dalibard, and others. Protection against lightning has become an important part of modern building construction.

In Germany Georg Lichtenberg, a writer and investigator in natural science took up the invention of Franklin and installed several lightning-conductors. He introduced the denotation "positive charge" for "resinuous" and "negative charge" for "vitreous" electricity. Lichtenberg was the first professor of Experimental Physics at the University of Göttingen. But he was also active in many other fields of natural science such as geodesy, geophysics, meteorology, astronomy, chemistry, even statistics and geometry. He is most well known, however, for his "Lichtenberg figures", produced from a mixture of sulphur and lead oxide in electric discharge. This is looked upon as the first xerographic recording process. Lichtenberg was the leading scientist in Germany of his time and his university lectures were much appreciated.

The next great contribution in electrostatics came with Coulomb and his law for attraction and repulsion between electrical charges. He also formulated a corresponding law for magnetic poles. The forces between two "point" charges are inversely proportional to the square of the distance between them. He also investigated the distribution of charges on electrified bodies, then observing that the charge on conducting bodies is limited to the surfaces. Coulombs results were used by Poisson to form a mathematical theory for electrostatics.

Charles Coulomb, born 1736 in Bordeaux, France, was both a physicist and an engineer, contributing to structural mechanics and friction. He remained a military officer for all his life, retired as a Lieutenant Colonel, and became member of the "Academie des Sciences".

After these fundamental works on electrostatic phenomena the modern development of electricity science started. Siméon Poisson, also a Frenchman born 1781. Inspired by Coulomb's work he developed a potential theory for electrostatics. The Poisson equation for the relationship between charge and potential plays an important role in mathematical physics. Before him, another Frenchman, Pierre Laplace, born in 1749, contributed much to mathematics and mathematical physics and formulated a differential equation that can be applied to a charge-free space. Laplace is considered as the founder of the theory of probability and his method of "Least square" is a powerful tool in statistical analysis. Poisson's distribution is wellknown in mathematical statistics. Poisson deduced the charge density on an ellipsoid and studied a system of two charged, separated spheres. The results of calculation on the spheres were in agreement with the experimental measurements by Coulomb.

The famous German mathematician Friedrich Gauss also contributed to the potential theory.

Michael Faraday, English chemist and physicist, one of the biggest contributors to the science of electricity, published in his "Experimental Researches in Electricity", investigations performed from 1831 to 1855. He is the father of the "triboelectric series". Elements and different materials were tabled in a series with those most prone to give positive charge on the top and those most prone to get negative charge at the bottom, when frictioned one material to another.

Two other Englishmen, James Clerk Maxwell and Olivier Heaviside gave the bases for the modern theory of electricity. In 1873 Maxwell presented a complete theory for the electrical phenomena as known by that time in his work "Treatise on Electricity and Magneticsm". Heaviside reformulated that work to make it better adaptable to practical application. Heaviside also developed an operational calculus system for electrical networks. An electrical atmospheric layer carries his name.

There are of course many new contributions to the theory of electricity according to new discoveries. For application to superconductors the theory of Bardeen-Cooper-Schrieffer (the BCS theory) and the London-London equations as well as the Josephson effect can be mentioned. N. Cooper also contributed to the theory of semiconductors.

Development of antennas, radiation of waves, are important for satellite communication today. The Swedish electro-physicist Erik Hallén made an early important contribution to the antenna theory.

1.1 Triboelectricity

1.1.1 Phenomenology

Electric charge is an old concept as phenomenon, but only by the modern atom physics it became possible to clarify its nature. According to experiments charged bodies are discharged by a current, i.e. transport of charges, if connected with earth through a conductor. A discharge can also occur through air (lightning spark). We have then the relation between "electrical current" and charges.

Conclusion: An electrical current is a transport of charges or charge carriers, i.e. charges in movement. The strength of current is defined as the passage per time unit of charges and depends on the character of the conductors. There are good and bad conductors. Electrical charges can be detected and measured through the mechanical forces they exert as described by Coulombs law, charges also represent energy e.g. in form of heat when passing through a conductor. Energy is also "stored" in a magnetic field around the conductor. The concept "potential" mentioned before is a measure of the energy of a unit charge. There are both positive and negative charges that can neutralize each others.

Definition: Static electricity is electrical charges at rest. They are recognized by the electrical field they establish between them and to non-charged conducting materials. The latter are "electrified" by induction (influence). Charge can be formed on all materials but are kept only if well insulated to the surroundings. The electric field exerts mechanical forces on charged particles in the field. This is described in the mathematical treatment of electrostatics.

1.1.2 The formation of static electricity

It was already pointed out that the electrification of materials is explained from the atomistic nature of matter. Both negative and positive charges are present in all materials. Usually they neutralize each others, but if separated charges are formed. This occurs either if two materials are rubbed against each others, or the field from an already charged body induces charge on a neutral one. These two procedures are described.

1.1.3 Triboelectricity

If two materials with different dielectric properties are in contact and then moved relative each others or suddenly separated a positive charge will appear at the surface of one material and negative on the other. The amount of charge produced depends on the compositions (atomic structure) of them and, in general, the rate of separating. The flatness of the surfaces has also an impact on the result. This kind of charge is called triboelectricity after the greek word tribein (= to rub).

The propensity of different material to get charged is usually represented in a triboelectric series as already mentioned. Table 1 is such an example. Those most prone to become positive are on the top and those negatively charged at the bottom of it. The distance between two materials gives an indication of the amount of charges formed when they are rubbed together. In the middle of the series are materials that remain almost neutral. Cotton is sometimes the division between positive and negative materials. The electric properties of cotton depends very much on humidity. Even metals can get charged!

Table 1. Triboelectric series

+	$\pm$	–
Air	Cotton	Orlon
Asbestos	Steel	Polyurethane
Glass	Wood	Polypropene
Mica	Amber	PVC
Human hair	Sealing-Wax	Silicon
Nylon	Nickel, Copper, Silver	Teflon (PTFE)
Wool	Gold, Platine	
Leather	Brass	
Lead	Ebonite	
Aluminium	Polyester	
Paper	Celluloid	

1.1.4 The nature of electricity

The atoms are the building-stones of matter. They comprise of a positively charged nuclear with negatively charged and almost weightless particles, electrons orbiting around them. The electrons are the elements of electricity, charge carriers, and the total charge of the electrons is equal to that of the nuclear charge making the atom electrically neutral. If the atom looses one or more electrons it is said to get "ionized". (see Fig. 1). Extra electrons can also be "catched", giving a negative ion The electrons of an atom possess both kinetic energy and because of their position relative to the nuclear, a potential energy. The sum of the two energy forms is normally constant. The electrons are moving in certain orbits belonging to so called electron shells. They can "jump" between different shells with energy change. If an electron is given sufficient additional energy it can leave from the atom, that will be positively charged ("ionized").

The different processes within an atom under energy exchange is more exactly described by quantum physics. The positive charge of the nuclear in the atom is

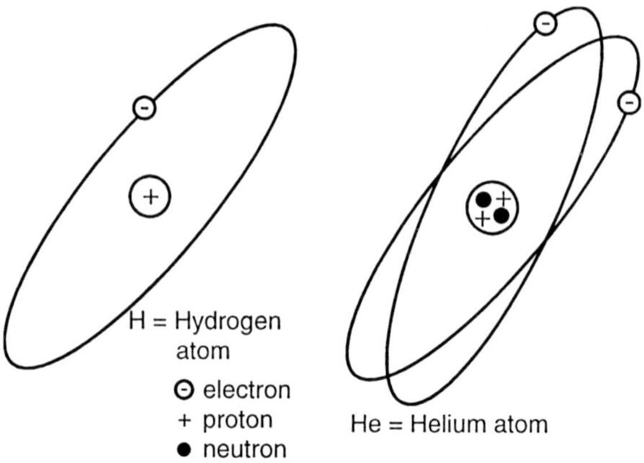

Fig. 1. Hydrogen atom with a positive nuclear, one proton, and two negative particles, electrons, orbiting around the nuclear. In the Helium atom the nuclear consists of two positive protons and two neutral particles with about the same weight as the proton. Two electrons are orbiting around the nuclear

characteristic for the element (compare with the periodic system). The electrons are more or less bound by the nuclear according to the outer electron configuration.

For a chemical compound the electron orbits are more complicated. This implies that some materials are more prone than others to release an electron, that can be caught by another material with stronger electron affinity. One of the two materials will be positively charged the other negatively (Fig. 2). This gives a physical picture for the formation of triboelectricity. A more precise account of the process claims knowledge about the quantum-mechanical relationships between the bonds in the two compounds.

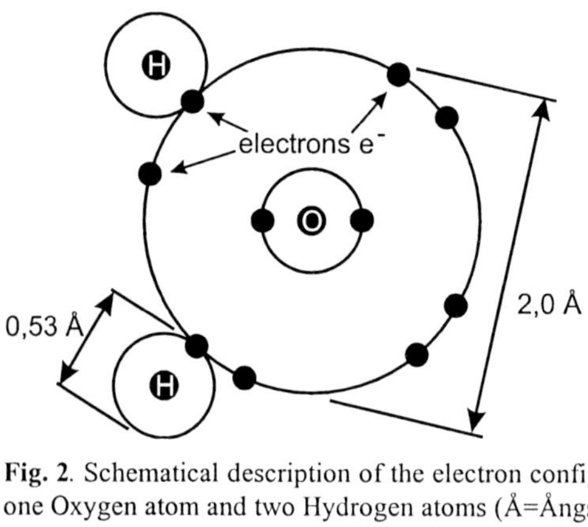

Fig. 2. Schematical description of the electron configuration in a compound, water, consisting of one Oxygen atom and two Hydrogen atoms (Å=Ångström, 1 Å $=10^{-10}$ m)

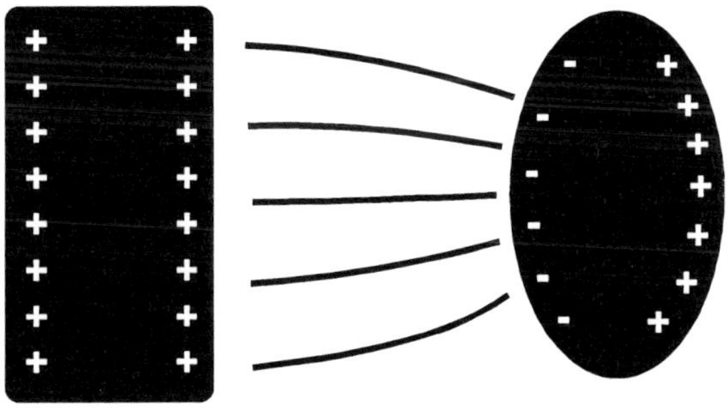

Fig. 3. Electrical induction (influence) between bodies. The positive charge on the body to the left has by influence separated the charge in the body to the right

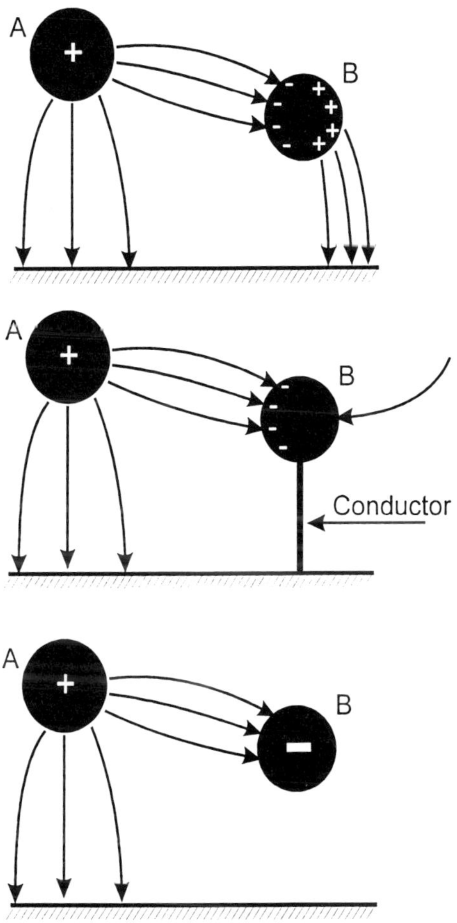

1.
Charged body A induced charges on the unearthed B

2.
Situation after connecting body B to earth. The field of body A holding the negative charges on body B.

3.
After breaking the connection to earth of body B, some negative charge is kept on body B.

Fig. 4. Induction (influence) between bodies and connection to earth

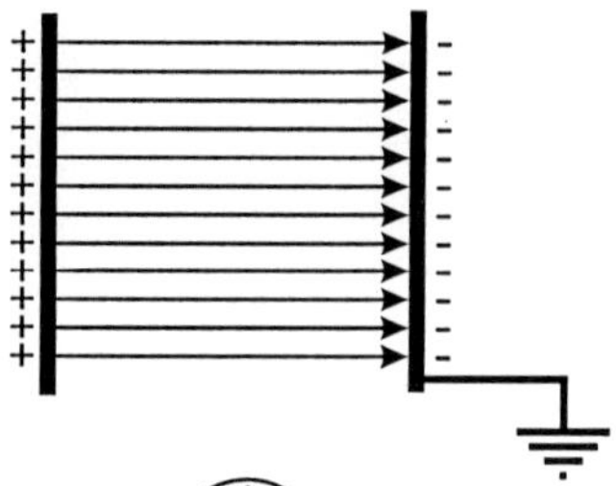

From physics it is known that an electrical field (e.g. in a capacitor) is defined to start at positive charges and stop at negative charges.

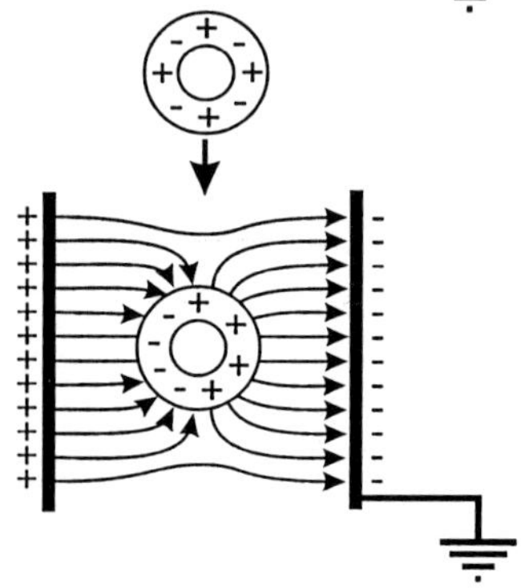

If a conducting body is introduced into an electrical field, the positive and negative charge of the body get separated.

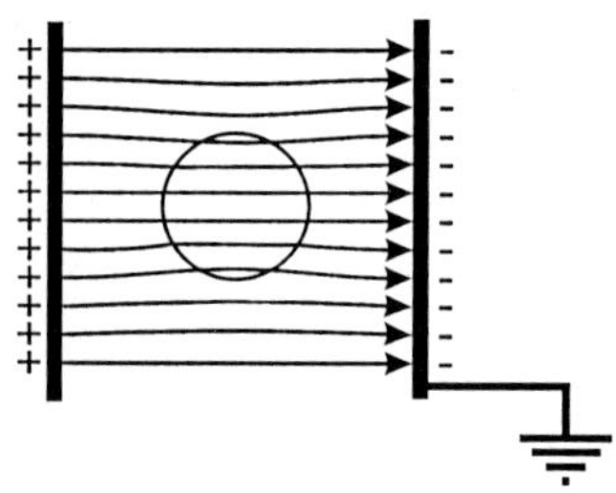

If instead a body with an insulating, dielectric material is introduced into an electrical field the electrical force lines are distorted but will penetrate the body that is polarized.

Fig. 5. Induction of conducting and non-conducting materials

1.2 The induced electricity ("influence")

An uncharged body near one that is charged can, if properly conducting, get its own charge separated in a positive and a negative layer (see Fig. 3 and 4). The charged body will attack the opposite charge on the neutral one and the equal charge will be repeated. For field configurations see Fig. 5.

If the polarized body is connected to earth, the repelled charge will leak off to earth. The body remains charged with the other polarity. A well known practical application is an electroscope. See Fig. 6, used to indicate charges.

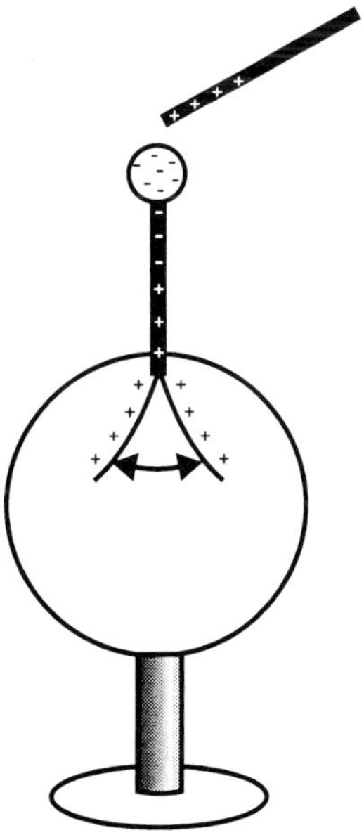

Fig. 6. Electroscope. When the rod is put near the upper globe charges are induced inside the electroscope

1.2.1 Electroscope

An electroscope consists in general of a glass container or similar entity, into which a conducting rod was introduced with two metal foils attached to its lower end. If, for instance, a triboelectrically electrified glass rod is moved near the upper part of the electroscope, charges are induced and the equally charged foils will separate (see Fig. 6).

1.2.2 Electrophorous

Alessandro Volta, most known for "Volta's pile", the first battery, also invented the electrophorous, Fig. 7. This was an arrangement for preservation of static electricity, characterized followingly: "electrified but once, briefly and moderately, never loses its electricity, and also repeatedly touched, obstinately preserves the strength of its signs". The principle of the electrophorous, based on electric induction, was given by Wilcke already in 1762. The following description is due to Wilhelm Becker (1919).

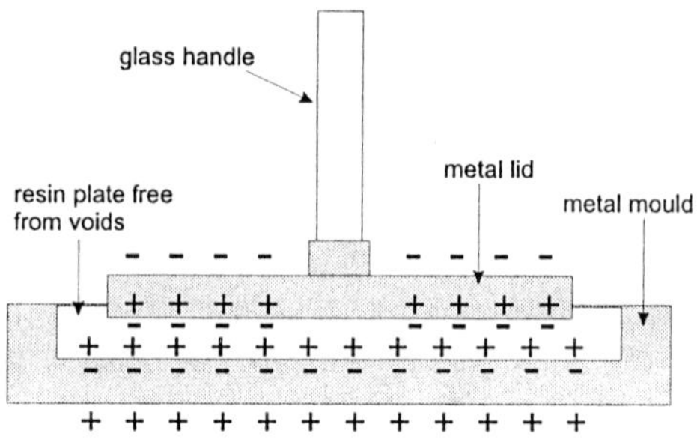

Fig. 7. Electrophorous, invented by A. Volta in 1776. Starting conditions after the placement of the metal lid upon the resin plate, according to W. Becker

By placing the resin plate into the metal form and whipping it with a "fox tail", the plate will get negatively charged. Placing the lid onto the resin plate and removing it again without touching the lid, nothing happens. The lid remains uncharged. Putting the lid back onto the resin cake and afterwards touching it with a finger, a spark of negative electricity is released. Removing the lid and touching it again with a finger a second spark is obtained, now of positive electricity. This procedure can be repeated almost infinitely without the resin plate losing any charge substantially. Only after a long time and frequently repeating the experiment, some charge disappears.

If we ask for the cause, why charges in the lid of opposite sign appear, it is first given, that these cannot have come from the resin plate by partition. The plate does not substantially lose any electricity in this case, and it also possesses only negative charge, while both kinds of electricity are appearing in the lid.

As the lid did not receive any electricity from the cake, it is natural to anticipate charges being present in the lid before, but neutralizing each others. The lid put on the plate its positive charge will be attracted by the negative charge in the plate and its negative charge repelled ("influence").

When then the lid was touched by a finger the negative charge was conducted to earth, the positive one kept by the negative charge of the plate. The lid being removed from the cake and a finger approaching it also the positive charge disappeared by a spark. If the negative resin plate is replaced by a glass plate, positively charged from rubbing with a piece of woolen cloth, the same phenomena are obtained but with the different charges appearing in reverse order.

The electrophorous being left by itself, it will keep its electric state for months. The explanation is that the positive and negative charges are mutually binding each other, thus obstructing loss of electricity. The surface of the resin plate is also rather effectively isolated from ambient air by the applied lid."

A. Volta established the relationship between the charge (Q), the capacitance (C) and the voltage (V) of a capacitor

$$Q = C \cdot V$$

The unit for potential, volt, is named after Volta. He also constructed an instrument for measuring very small charges and studied atmospheric electricity.

1.3 Charge, potential and field

1.3.1 General conditions in an electric field

For a long time, the science of electro-statics played a role of demonstration for electrical phenomena beside electro-dynamics. Technical applications have, however, improved the interest in electrostatics. Examples are, for instance, filters for purification of smoke from factories, where an electric field will eliminate charged particles and electrets (charged plastic foils) used as membranes in high fidelity microphones. The most important reason for the activation of electrostatic investigations is, however, the advent of synthetic (plastic) materials and their wide use in modern society. These materials are often easily charged, positively or negatively. Discharges of them can create annoyances in many forms: explosions in factories, destroying or disturbing electronic equipment and causing fire.

An almost new science for counteraction of these effects during the last three decades: The science of "antistatics" was born. This is very important for the modern electronic industry, using a great lot of sensitive components and equipment and in the fast developing information technology (IT) as a whole. This has become a great concern for the electronics industry. All packaging and protection materials used for components and equipments must be antistatic. How that is achieved is the main goal of the book.

1.3.2 Mathematical-physical relations in electrostatics

The description in the following chapters are comprehensive without an advanced knowledge of the mathematical formulation of the theory of electricity. Basic concepts and relations as to electrostatics will be given here.

The unit of charge is named "coulomb" (after Coulomb) or, in the MKSA system, "ampere seconds". Potential and voltage (=difference between two potentials) have the unit "volt". The "dimension" of volt is deduced from energy and charge (see further on). Capacitance, the ability to collect charge in a capacitor, has the unit farad as defined in Volta's law: $Q = C \cdot V$.

In this context energy, W, is defined as:

$$W = \frac{Q \cdot V}{2} = \frac{V^2 C}{2}$$

The unit of W in MKSA is nm = "newton meter", where "newton" is the unit of (mechanical) force. Electric field E is "voltage per meter", V/m. In a plan parallel capacitor: E=V/d, d = distance between electrodes. E is a different concept than W, Q, V, C, having not only a magnitude but also a direction.

Electric "field lines" are defined to begin at positive charges and end at negative.

The field strength was given as V/d for a capacitor. Generally it is the force the field exerts on a particle with unit charge. The force is proportional to the charge: $F = E \cdot Q$. The unit of the force is "Newton" in MKSA.

As seen from the field lines are parallel and the field strength is constant everywhere =V/d and, consequently, the force on a unit charge will be constant in the field. The "work" performed to move such a particle from the positive to the negative electrode will be

$$F \cdot d = \frac{V}{d} d = V = V_1 - V_2 \quad (V_1 > V_2)$$

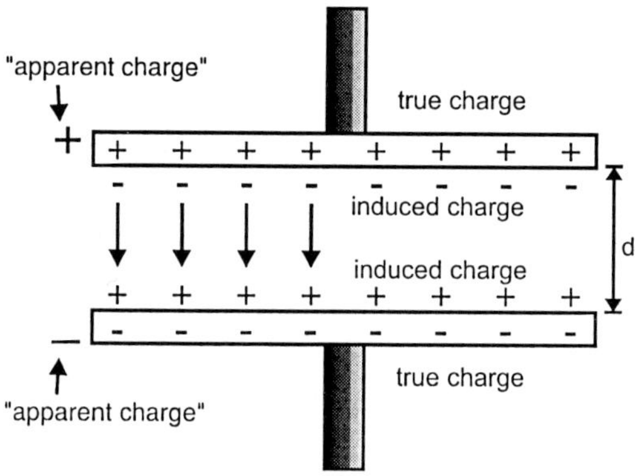

Fig. 8. Electrical conditions in a plan parallel capacitor. The term "apparent charge" is not used in modern textbooks on electricity, but it is a useful, intuitive concept.

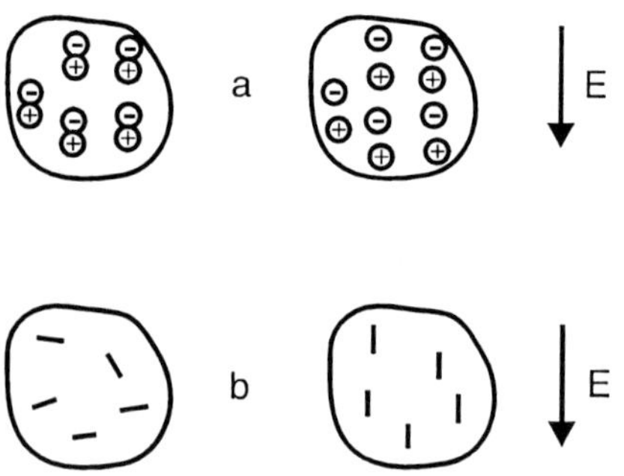

Fig. 9. Examples of charge displacement **(a)** and polarization **(b)** in an electric field

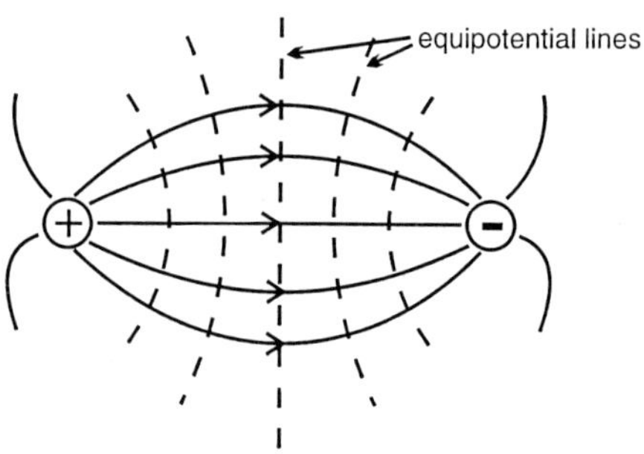

Fig. 10. Electric field between charges

Thus the potential difference, i.e. voltage, is equal to the work performed. In general a potential V is introduced for description of an electric field (see further on) and the field strength will thus be

$$E = -\frac{dV}{dr}$$

r = geometrical length of field lines

The minus sign is due to the fact that the potential will drop in the direction of the field (= the direction of the force exerted on positive charges).

If a dielectric material (dielectrics) is introduced between the two electrodes of a plan parallel capacitor (see Fig. 8), the field "polarizes" (see Fig. 8) the dielectrics and thus increases the field strength. Polarization implies separation of positive and negative charge or aligning existing dipoles in matter. An induced charge density is the result (Fig. 8). This depends on the field strength and polarizing properties of matter. The polarization, P, is in general proportional to the field strength E and a new concept, electric displacement D, is introduced:

$$D = \varepsilon_0 E + P$$

ε_0 is a system constant

Applying $P = k \cdot \varepsilon_0 E$, where k is a constant, we get:

$$D = \varepsilon_0 E + k \cdot \varepsilon_0 \cdot E = (1 + k) \cdot \varepsilon_0 \cdot E = \varepsilon \cdot \varepsilon_0 \cdot E$$

The factor ε is called the dielectric constant of the dielectrics used.

There is a relation between the charge density q on the capacitor electrodes and D

$$D = q$$

and over a surface of A m^2 with charges this gives, according to Gauss

$$\sum Q = \iint D dA$$

where Q and D vary over the surface. D is also called electric flux density.

D, like E, has a direction. Such quantities are called vectors and it is the component of the vector perpendicular to the surface A that is used in the integral, often denoted D_n.

From the above we know that there is a relationship of charge (Q), capacitance (C) and potential (V). The difference V_1-V_2 between two potentials is called voltage. If the voltage between V_1 and a potential $V_2 = 0$ is found, the potential V_1 is also given. We recollect that

$$V_1 - V_2 = -\int_{\bar{r}_1}^{\bar{r}_2} E\, dr$$

where $\bar{r}_1$, $\bar{r}_2$ means starting and ending points for the line integration.

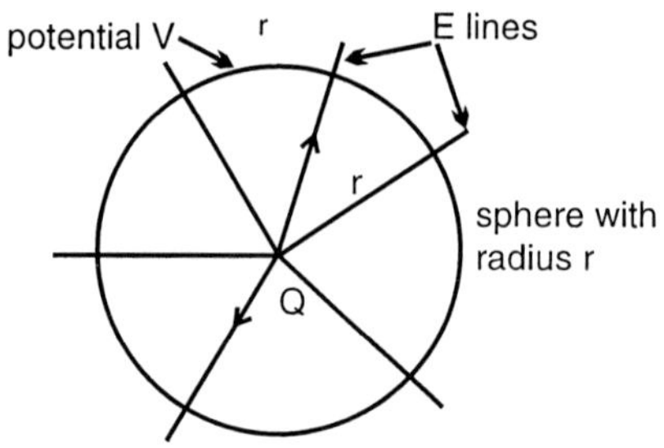

Fig. 11 The radial field from a point charge.

From a point charge Q with large distance to the surroundings, the electric field will be almost radial (see Fig. 11).

Because of the symmetry, E will be constant over the spheric surface $=4\pi r^2$. Using $D = D_n = \varepsilon_0 E$ ($\varepsilon\varepsilon_0 E$ a dielectric space). Thus the charge density $q = \varepsilon_0 E$ and

$$Q = 4\pi r^2 \varepsilon_0 E$$

If V_r, the potential on the sphere surface, and V_∞, potential at $r = \infty$, we have

$$V_r - V_\infty = \int_r^\infty E\,dr = \frac{-Q}{4\pi\varepsilon_0}\left[\frac{1}{r}\right]_r^\infty = \frac{Q}{4\pi\varepsilon_0 r}$$

Setting $V_\infty = 0$, then $V_r = \dfrac{Q}{4\pi\varepsilon_0 r}$ is the potential distribution around Q.

The capacitance, C, is generally defined as:

$$C = \frac{Q}{V}$$

Thus if an isolated charge Q is regarded as a small sphere of radius a we can write

$$V_a = \frac{Q}{4\pi\varepsilon_0 a} \text{ and } C = 4\pi\varepsilon_0 a$$

The system constant is ascribed the dimension:

$$\varepsilon_0 = \frac{10^{-9}}{36\pi} \approx 8.8542 \cdot 10^{-12} \frac{\text{Farad}}{\text{m}}$$

For air the dielectric constant $\varepsilon \approx 1$.

The original deduction of relationships in an electric field starts with Coulomb's law for forces between charges

$$F = \frac{Q_1 Q_2}{4\pi\varepsilon_0 r^2}$$

setting $Q_2 = 1$ we get

$$E = F = \frac{Q_1}{4\pi\varepsilon_0 r^2}$$

The potential V from several charges in a space is received as the superposition of potentials from single charges, for instance

$$V = \frac{1}{4\pi\varepsilon_0}\left[\frac{Q_1}{r_1} + \frac{Q_2}{r_2} + \frac{Q_3}{r_3}\right]$$

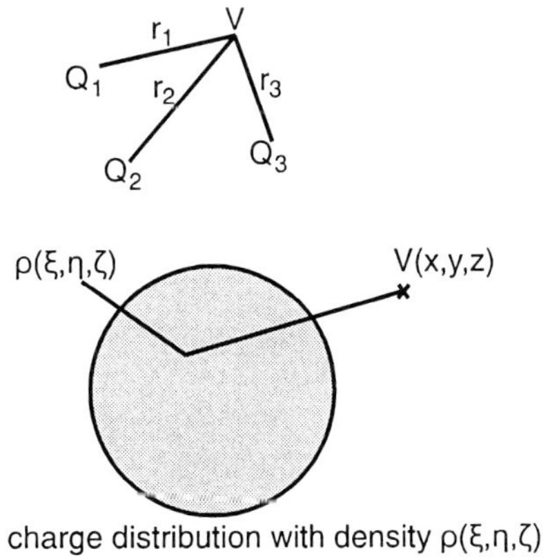

Fig. 12. Discrete and distributed charges.

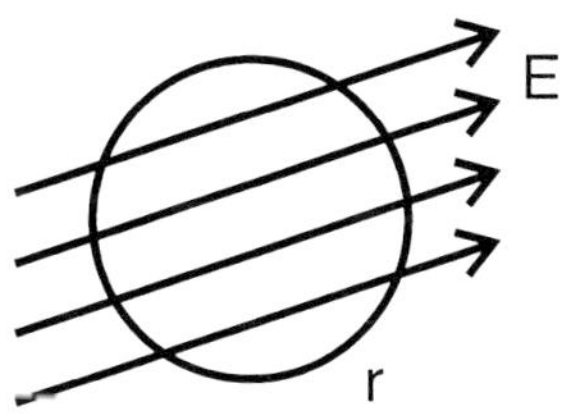

Fig. 13. Conservative field

Potential from a charge distribution

$$V(x,y,z) = \frac{1}{4\pi\varepsilon_0}\iiint\frac{\rho(\xi,\eta,\zeta)\,d\xi d\eta d\zeta}{\sqrt{(x-\xi)^2+(y-\eta)^2+(z-\zeta)^2}}$$

In a charge-free space the potential V satisfies the Laplace equation:

$$\frac{\partial^2 V}{\partial x^2} + \frac{\partial V}{\partial y^2} + \frac{\partial V}{\partial z^2} = 0$$

and in a space with charges, the Poisson equation:

$$\frac{\partial V^2}{\partial x^2} + \frac{\partial V}{\partial y^2} + \frac{\partial V}{\partial z^2} = \frac{-\rho}{\varepsilon_0}$$

The electric field in a charge-free space is said to be conservative, no energy change, that is a closed line integral will be zero (Fig. 13).

$$\oint E\,dr = 0$$

When an electron, charge e, is moving in an electric field it will gain the energy:

$$dW = -e \cdot (V_1 - V_2) \quad (V_1 > V_2)$$

The energy gain for V_1-V_2=-1 is called an "electron volt", a common energy unit in physics.

In cases with charges near conducting walls, some methods can be used to depict the electric field:

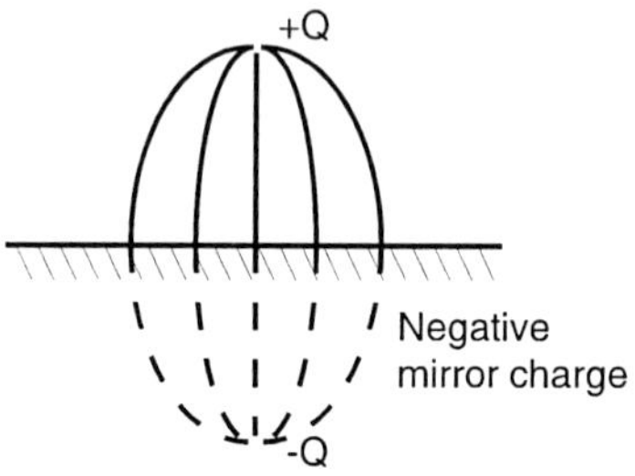

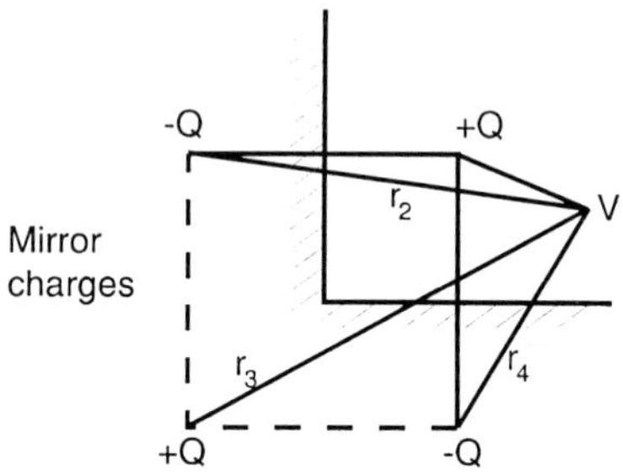

Fig. 14. Charge above conducting surface **Fig. 15.** Charge in corner of two conducting surfaces

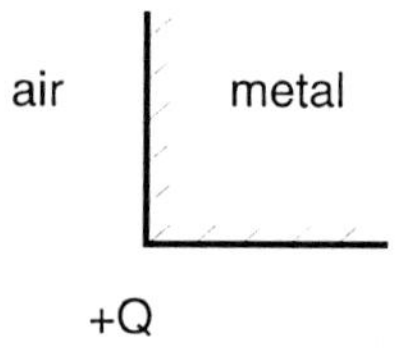

Fig. 16. Charge outside wall with corner

The basic rule is to reflect all charges in all planes, fig. 14 and 15. But in a case like the next one (fig. 16), conformal mapping is to be used.

The electric field is always perpendicular to a conducting surface. The potential over a charged insulated metal body is constant (E=0), but charge distribution varies. For a charged insulator see fig. 17 and 18. Fig. 19 and 20 show charged metallic bodies.

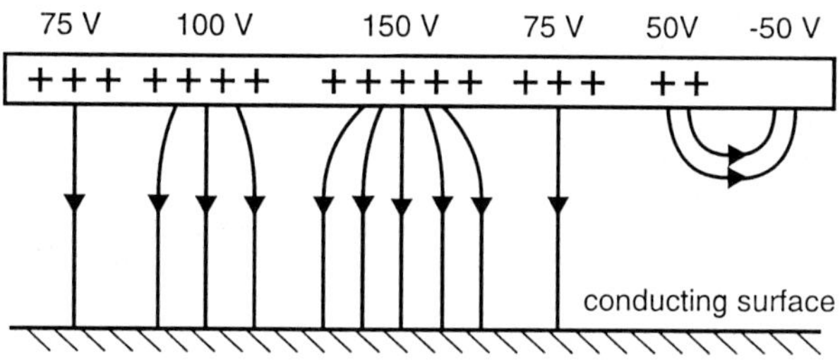

Fig. 17. Charged insulator. As the charges are bound and cannot easily move, the potential can vary over the surface. Owing to the polarity of the charges, the field lines will show different courses.

Finally the relation between charge, Q, and current, I, is quoted

$$I = \pm \frac{\partial Q}{\partial t}$$

plus sign for increasing and minus for decreasing charge.

1.3.3 Conductors, semiconductors, insulators

Originally the ability of different materials to conduct a current was characterized as "conductors" and "insulators". Metals are typical conductors, ceramics and plastics typical insulators.

Early it was discovered that many liquids - in first case water with a dissolved chemical compound form ions, i.e. is divided in a negative and a positive part - can conduct a current. The conductivity varies very much for both conductors and insulators.

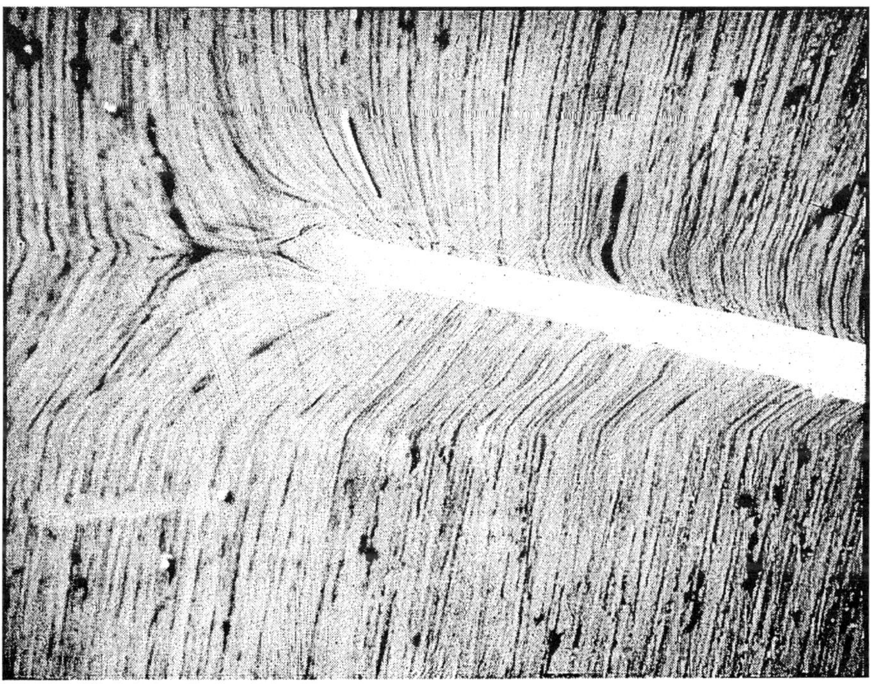

Fig. 18. A hair is a good insulator. This picture is taken of a hair in a SEM (Scanning Electron Microscope), the pressure being 10^{-6} mm Hg. The hair is charged by the electron beam from the microscope. Its resistance is about $2 \cdot 10^{14}$ ohms, a good insulator! The charge will not leak off and the electric field from the hair charge will distort, as seen, the pattern of the underlying brass plate. Observe the peak effect at the hair top, field concentration!

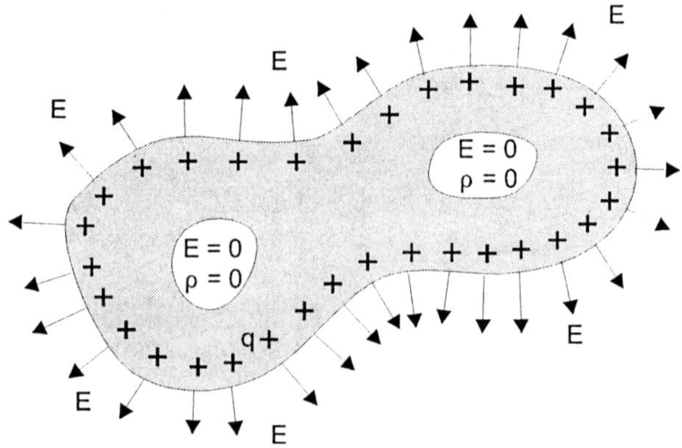

Fig. 19. Charged metal body with cavities. E = field strength,ρ = volume charge density, q = surface charge density.

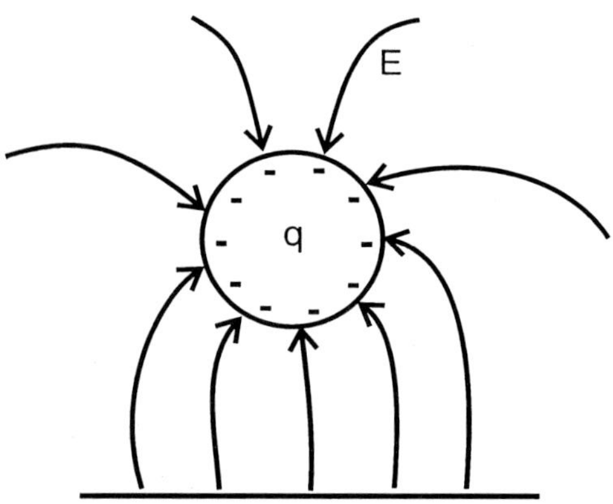

Fig. 20. Charged metallic sphere. E = field strength q = surface charge

Materials with conductivity mainly in the range between conductors and insulators are known as "semiconductors". Their conductivity depends, however, on admixtures of special elements. The materials are said to be "doped" by them. They are used as "active" elements in electrical circuits, like electron tubes. Undoped semiconductors are called intrinsic semiconductors and the conductivity varies only with temperature.

The conductivity is determined from the electron structure of the matter, current generally achieved by electron transport. In semiconductors there are two charge carriers electrons and positive "holes". The latter are atoms who lost one or more electrons. By exchanging electron with a neighboring atom a hole transport takes place. In solid state physics the electrons of semiconductors are said to belong to a "conducting band" and the hole to a "valence band". These are electrically separated by a "band gap" that depends on the electron structure (bonds) in the material. This gap must be overcome for an electron to move from the valence to the conduction band. The process is called electron - hole - pair creation as the electron will leave a "hole" in the valence band and add an electron to the conduction band (EHP).

New forms of conducting materials have been achieved by admixtures of conducting elements as carbon and aluminum in otherwise poor conducting plastics. Intrinsic conducting in some plastics as well as conduction by doping of them was discovered. All these admixtures play an important role for the conductivity.

The following table illustrates the variances in the conductivity of different materials.

Table 2. Resistivity for different materials

Material	Resistivity (ohmmeter)
Silver	16.10^{-9}
Copper	17.10^{-9}
Gold	23.10^{-9}
Aluminium	27.10^{-9}
Charbon (filement)	40.10^{-6}
1 n HCL-solution	10^{-2}
Germanium	46.10^{-2}
Bacelite	10^{5}
Pertinax	10^{8}
Celluloide	2.10^{8}
Wood	3.10^{8}
Plexiglass	$> 10^{10}$
Glass	$10^{11} - 10^{12}$
Mica	$10^{13} - 10^{15}$
Paraffine	10^{15}
Amber	$> 10^{16}$
Polystyrene	10^{17}

1.3.4 Antistatic materials

With antistatic materials was earlier meant materials with a specific conductivity range. This denotation has, however, become to cover all kind of materials with the property to reduce the formation of charges and dissipate any charge within a few seconds. Some of them also have the special property to shield against electrical fields. They are not penetrable for fields. Antistatic materials are usually characterized by their surface conductivity. Static dissipation was before used for materials in one conductivity range, but is now in another, the main range of antistatic materials between insulating and conductivity materials. Figure 21 gives a survey of different scales for the conductivity properties of matter.

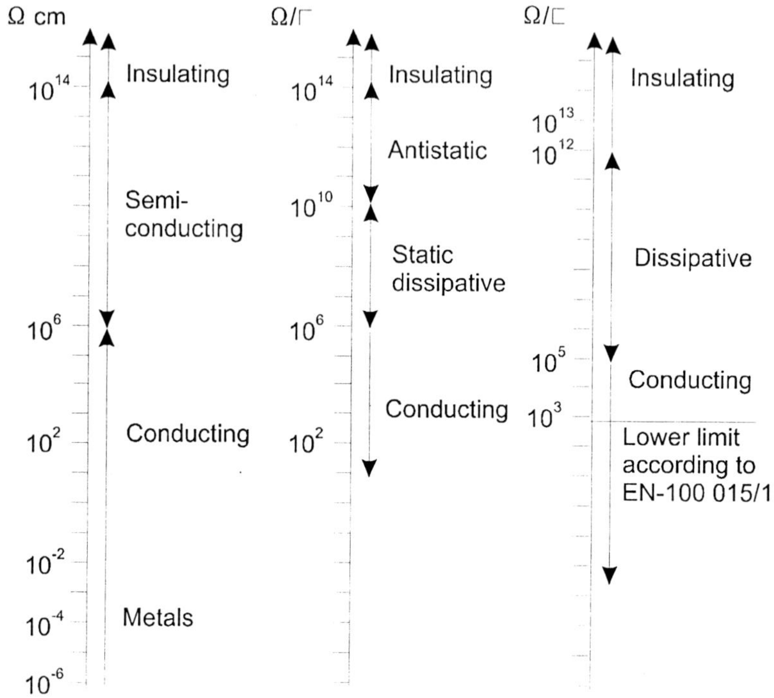

Fig. 21. Scales used for the resistivity of matter. To the left an old, physical, division for resistivity properties is found. In the middle the former electrostatic description of resistivity is given. To the right is the scale of resistivity valid according to the European standard EN 100 015/1.

2 Relation to electronics

In the first chapter was shown how static electricity, i.e. electric charges can be created. These charges are transferable by different ways to electronic components.

2.1 Charging

2.1.1 Influence from environment

The formation of charges is highly dependent on the relative humidity (RH-value). This is not due to improved conductivity of the air at higher relative humidity, as could be thought, but to the formation of a thin water layer upon otherwise insulating surfaces. See fig. 22

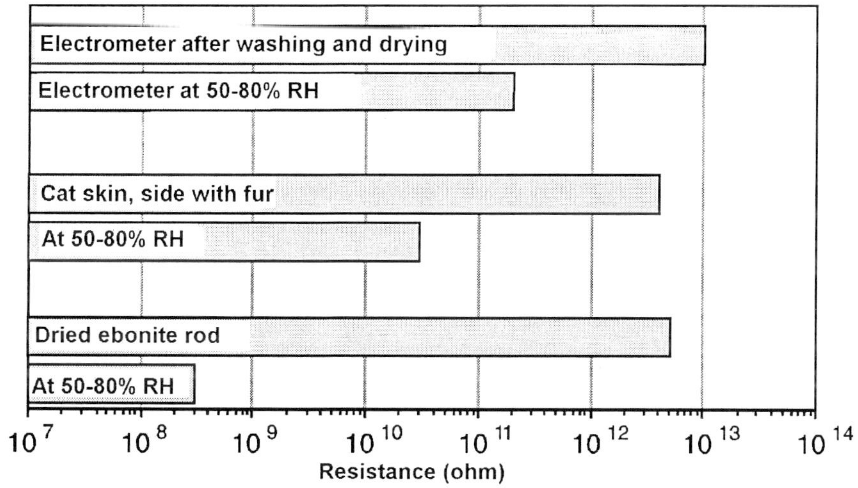

Fig. 22. Insulating resistances of materials used to demonstrate electronic phenomena, when stored at room temp. and RH, summer season, and cleaned and dried. The diagram explains why electrostatic experiments not always are successful! Comparison made at increased RH-values

Pure water and pure air are efficient insulators but are able to contain impurities (ion generators) this implies that water on the surface of a material counteracts the charging process.

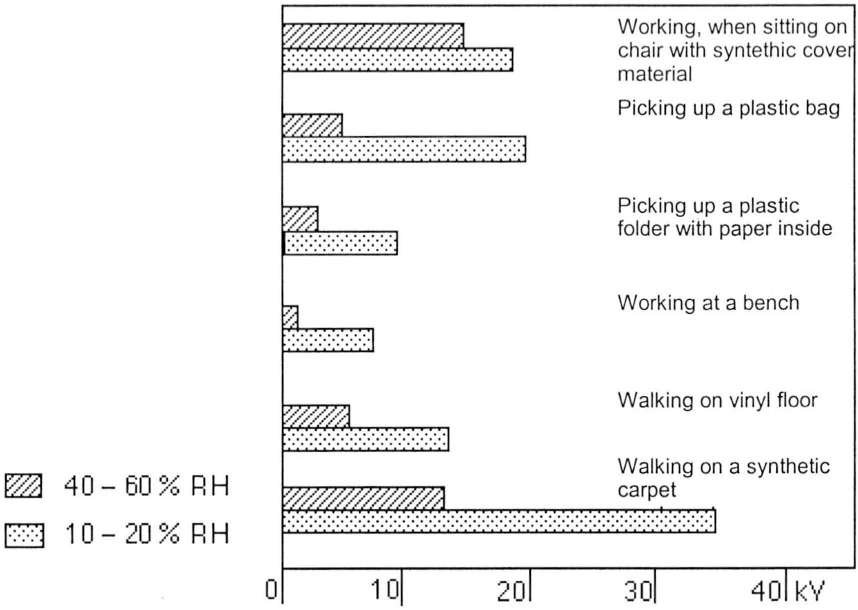

Fig. 23. Charging at different situations

Certain materials have a higher propensity to absorb moisture than others. Cotton is rather safe against being charged already at 30-40% RH, while nylon is safe only at 50%. The importance of the relative humidity can be studied in fig. 23 the higher the humidity the lower the charging. Fig. 24 shows RH-variation.

The relative humidity is a measure of the amount of water that a certain volume of air (or other gas mixture) contains at a certain temperature compared with the maximum amount of water possible to be absorbed at that temperature. This means that cold air taken into a building and warmed up improves the ability to absorb water, i.e. the RH-value decreases if no other water is present to be absorbed.

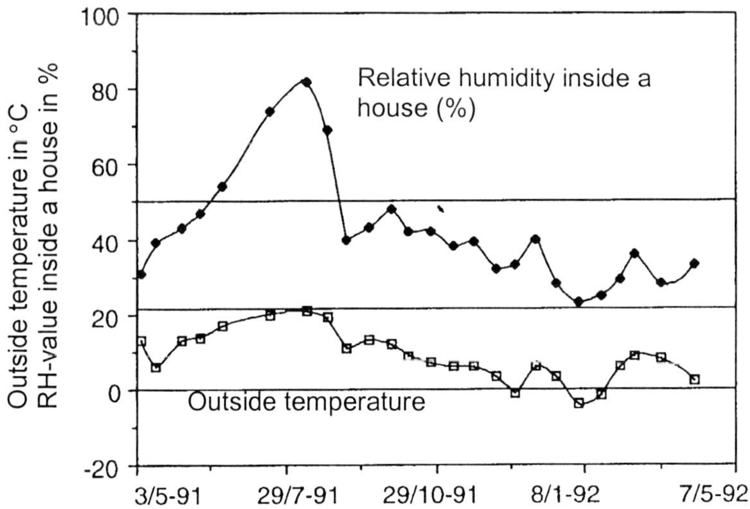

Fig. 24. Relative humidity inside a house and outside temperature measured during the period May 1991 to May 1992 in the area of Stockholm

At winter season the RH-value is often 80-100%, but the temperature being perhaps -10°C. When the air is taken into a building and warmed up to +25°C the RH-value goes down to about 7%! The air is experienced as dry and the creation of electric charges is favoured, as all materials at such conditions get very dry. The result will be a strong increase of surface resistivity preventing the charges to leak off.

The relative humidity can be found from a so called Molliérdiagram, see fig. 25

$$\frac{RH_i}{T_i}\exp(-5370/T_i) = \frac{RH_u}{T_u}\exp(-5370/T_u)$$

where RH_i and RH_u = RH-values inside and outside a house, T_i and T_u are corresponding temperature values.

Example:

If RH_u = 72%, T_i = 22°C and T_u = 3,7°C we recieve

$$RH_i = 72\frac{(273+22)}{(73+37)}\exp\left[5370\left(\frac{1}{295}-\frac{1}{277}\right)\right] = 23.5\%$$

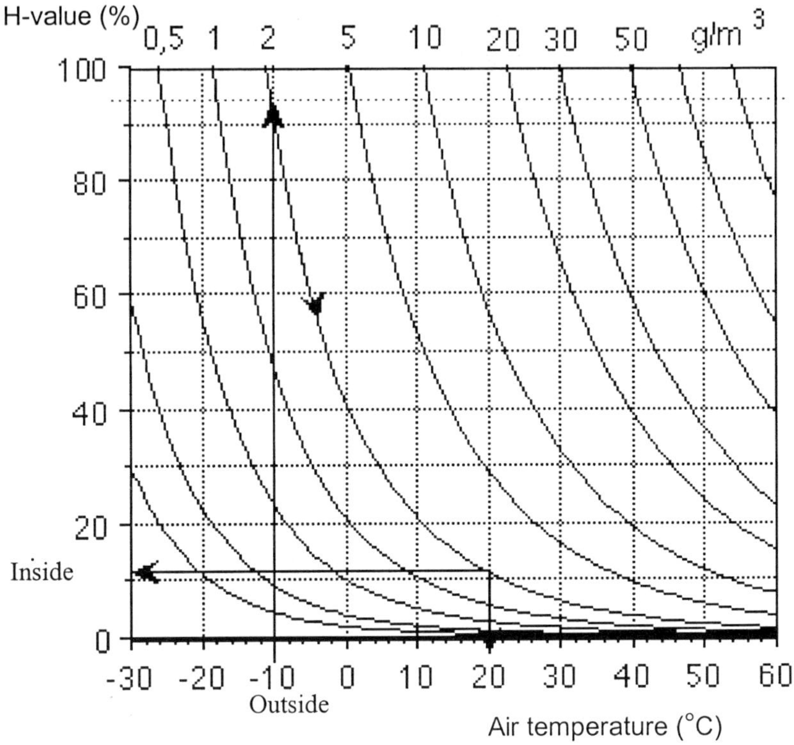

Fig. 25. Molliérdiagram. From such one it is possible to determine how the state parameters (temperature and RH-value) would change if an air mass is transferred from one state to another. Using the starting conditions a curve in the diagram is followed to the other state and the new parameters are recorded. In the diagram is given how the values of a wintertime day (-10°C and 95% RH) are changed if the air is sucked into a house with +25°C. RH is read to be about 12%. In a real case other factors, as plants and people will raise the RH-value above this value

2.1.2 Contact between materials

As already shown bodies become charged when one material surface get rubbed against another. A common conception is that formation of charges results from the friction process, but triboelectricity comes about when the two surfaces are loosing contact. A contact can be established with or without relative movement of the surfaces. At increased movement the contact surface becomes larger with higher friction. Raising the contact pressure gives the same result, because surface inequalities are eliminated. All these actions will produce more charges after separation of the two surfaces. See fig. 26.

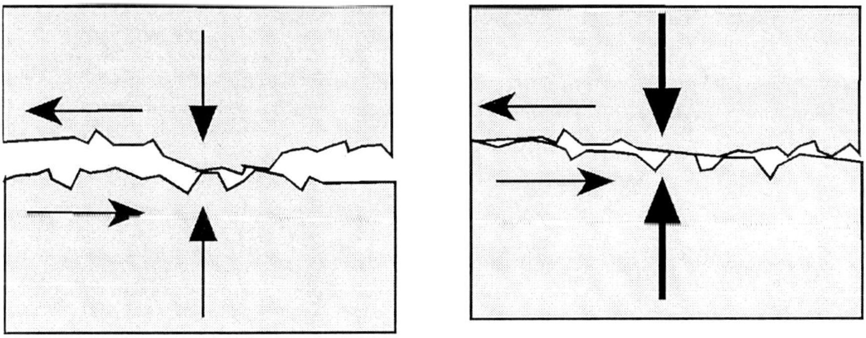

Fig. 26. Contact surface enlarging by strong pressure. Different surface structures will influence on the formation of surface potentials and traps

In general, the charging effect increases with the relative speed of the surfaces in contact. But other parameters must be considered. The friction heat is increased with the speed changing, mostly decreasing, the charging ability. A short contact period admits less discharge thus raising the charging effect. Because of these phenomena a prediction of the final charge will be difficult.

Experiments were performed with tapes applied at an electroscope. They were removed with the same force but different speed. Charges left on the electroscope gave rise to a deflection by the instrument. No significant difference between the trials could be observed. The speed had no influence!

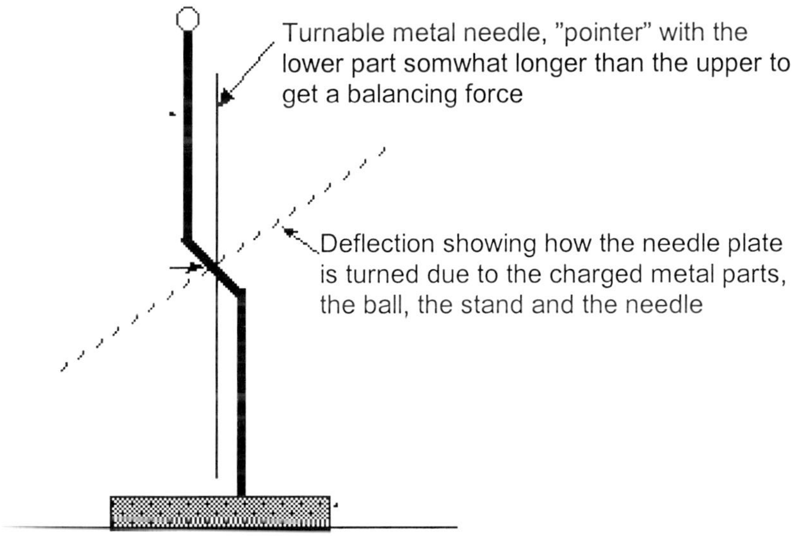

Fig. 27. A simple electroscope. The suspended bar makes for small charges deflections proportional to their amount

After rubbing and separating two bodies the surfaces will remain charged, positive charge on one of them and negative at the other. This process is called "charge separation" in contrast to the otherwise neutral state. Should a conductor be used as one part and connected to "earth" its charge will disappear that rapidly not to be observed.

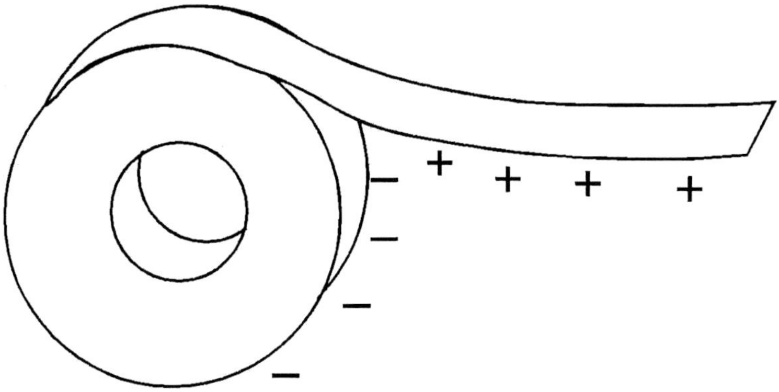

Fig. 28. If plastic tape on a roller is unrolled charges can be separated, proving that a certain material will be triboelectrically charged with both kind of electricity, only the friction is sufficient

In fig. 27 is shown an electroscope that is a common instrument to indicate charges.

Common charge generators are found in table 3.

A. Working surfaces
- Polished wooden surfaces
- Synthetic mats
- Unearthed metal plates
- Glass

B. Floors
- Vinyl plastic
- All waxed surfaces

C. Chairs
- Fibre glass
- Plastic
- Vinyl covered
- Unearthed metal
- Polished wooden

D. Personnel clothing
- Inside clean rooms
  - Synthetic frocks
  - Synthetic shoes and gloves
- Private clothes
  - Woollen
  - Synthetic materials
  - Silk
  - Jersey

E. Packing materials for parts and equipments
- Polyethen bags and foils
- Polyethen bubble and foam plastics
- Poly-urethane packing flaws

F. Cleaning and test areas
- Fast gas flows
- Temperature chambers
- Drying rooms

Table 3. Common charge generators

2.2
Direct electric discharges

A discharge that gives rise to an electric current path through for instance, a component or part of electric equipment is called a direct electric discharge. An air gap can be a shorter range of this path. See fig. 29.

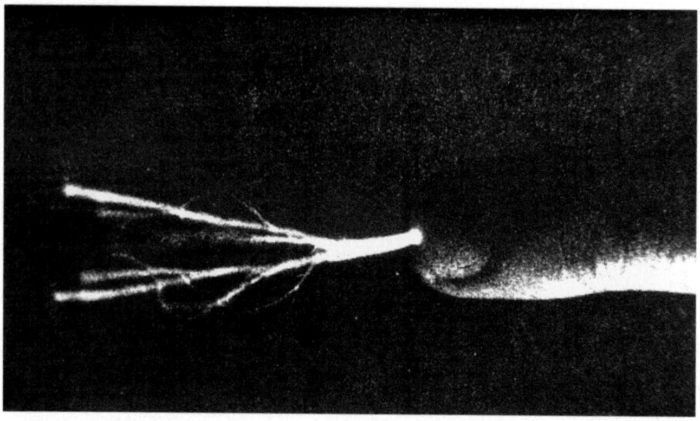

Fig. 29. A negatively charged teflon plate is discharged to a finger tip. The finger was moved near the plate in total darkness. It looks like the discharge is going from the finger to the plate, but the reverse happened. A corona (light effect) is to be observed near the fingertip. The total length of the lightning is about 55 mm, the voltage should have been about 60 kV, since a free air discharge was established. Compare this figure with the computer processed picture of figure 88, p. 90, they have the same structure!

2.2.1
Charge from a person transferred to a component pin

If a charged object or person touch a component pin a destructive discharge can occur. The first ESD damages observed originated from the operators, handling components and parts, which resulted in many fundamental investigations on ESD suscecptibility. Standardized testing models and equipments were elaborated. Possible potentials in an environment not ESD-protected are shown in figure 23.

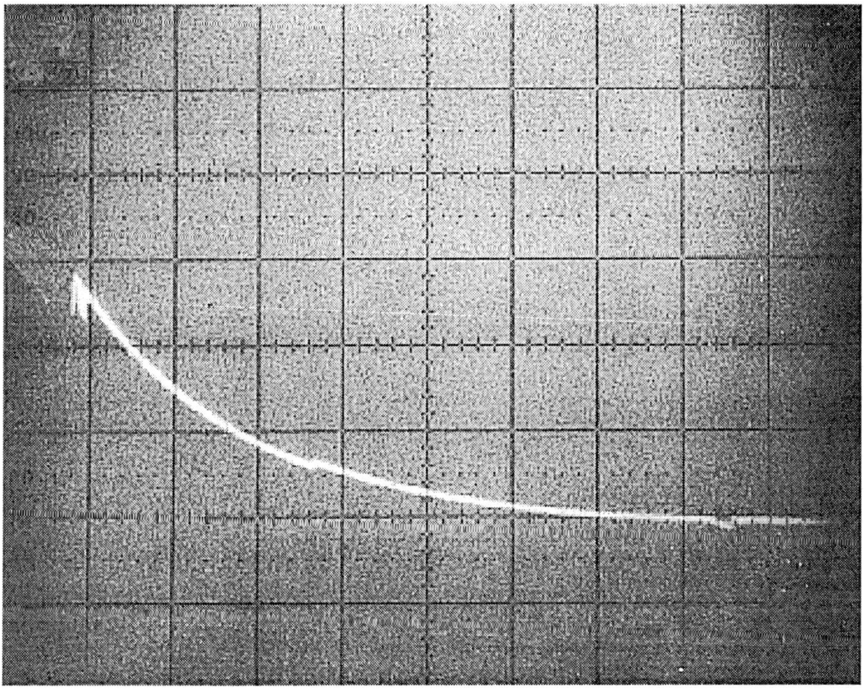

Fig. 30. Discharge pulse from a person through a high voltage probe
Y-axis: 1000V/square
X-axis: 5 ms/square

Testing of the propensity of protective garments to get charged was performed followingly. A number of test persons rubbed their bodies against a textile clothed chair when sitting for just a couple of seconds, rose to their feets and touched their fingers to a high voltage test probe with 100 MΩ input resistance. See fig. 30 and 31.

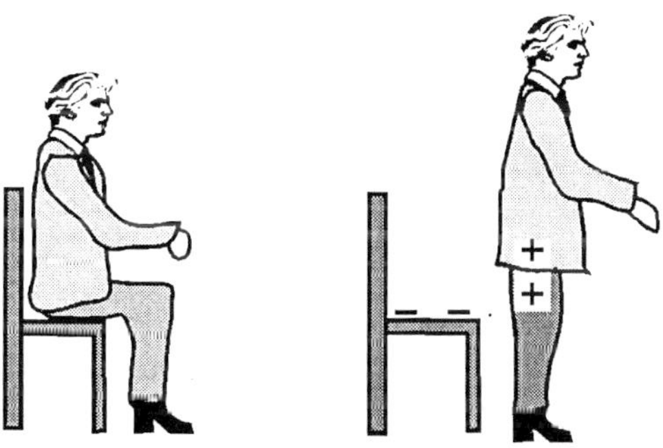

Fig. 31. Charging of people when separated from a chair

Regarding a persons body as a capacitor and his arm as a resistor of resistance far less than 100 MΩ, the following deduction is possible.

The time of capacitor discharge through a resistor is denoted τ, defined as the time elapsed from start to the point of time when the capacitor voltage is lowered to 1/e of its peak value (e = natural logarithim base).

τ=RC, R=100 MΩ, τ estimated from the figure to 10-23 ms and inserted:

$$C = \frac{\tau}{R} = \frac{10^{-2}}{10^{8}} = 100pF$$

This capacitor value is in agreement with that used for ESD test simulators for discharge from operators. See fig. 32 for an empirical ESD-pulse.

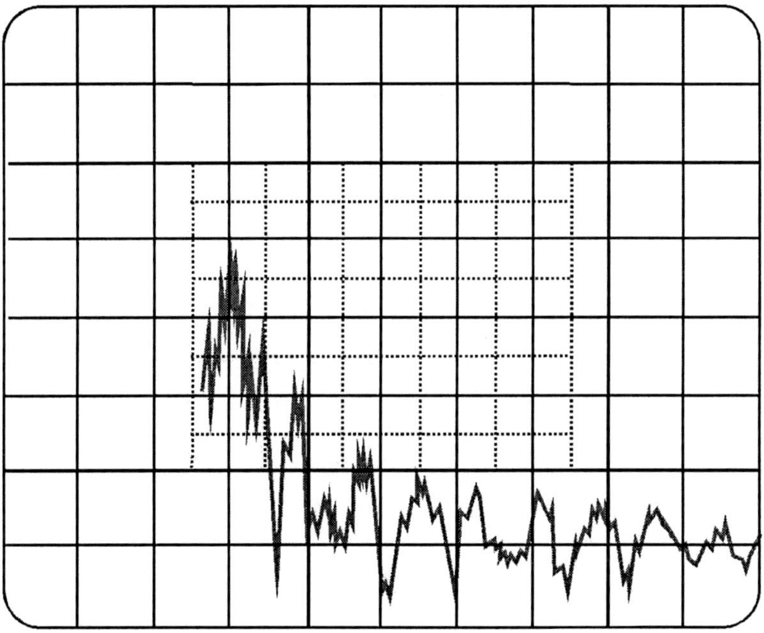

Fig. 32. Empiricial ESD-discharge through a component. The smoth "e-function" generated by ESD-test generators is here replaced by a more real oscillating discharge

2.3 Self-charging

A component, e.g. microcircuit, get charged for instance when sliding through a test equipment or component tube or wiped by a flush of dry air. The charged formed on the encapsulation will induce charges at the component pins. If these are touched by conducting objects a discharge will occur. Charges are induced on all metal parts of the component. The "corner" pins are most likely to be touched. It is difficult to protect components against this kind of discharge.

2.4 Charging by induction (influence)

A component can also be charged by induction from outside (external electrical field). A charged body passing the component is able to induce charges on it. Inside metal parts can get charged why it is also in this case difficult to protect against destructive discharges. See fig. 33.

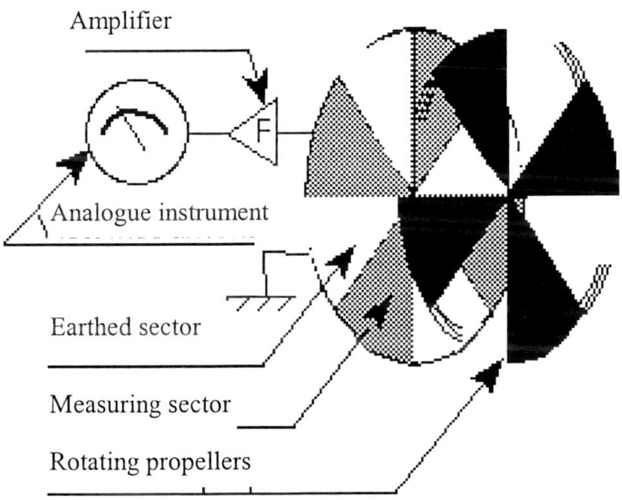

Fig. 33. Schematic picture of a "field mill". This is an instrument for measuring electrostatic field strength based on the induction principle. The measuring unit is a round plate divided into sectors. Every two sector is connected to earth and the others serve as capacitive sonds connected to a measuring instrument. In front of the measuring plate is applied a rotating propeller whose "blades" have the same shape as the sectors of the plate. Thus the measuring sonds are periodically exposed to the electrical field to be measured. The amount of charge induced on the sonds is proportional to the field strength, the unit of which is V/m

The field strength is essentially dependent on the charge density (As/m^2) or voltage of the body, the distance between the mill and the body and the measuring configuration. If field strength, distance and geometry are known the charge density can be calculated too. Fig. 34 shows a measuring instrument.

Fig. 34. A commercial field mill

2.5 Rolling gates as ESD-generators

An often overlooked reason of ESD-damages is the induction of charges on a component from another charged object. In industry so called rolling gates are often used. They can be made of a synthetic stuff in sheet form, that is rolled up for the gate to be open and unrolled to close it. In the unrolling process the sheet is seperated from other parts of itself. Evidently a condition for generating triboelectricity at normal or low air RH-values! This can result in voltages of the sheet as high as 100 kV, depending on the RH-value. See fig. 35.

The field itself can be the reason of break down in components in the neighbourhood containing thin field oxide layers. Metal objects can submit charge separation, resulting in strong discharge currents when earthed. Free cables may of the same reason have a first discharge when connected to an equipment. If disconnected again and moved out of the field space, they are still charged and when connected again to another equipment outside the field a second discharge will take place but now with charges of opposite polarity!

The pulse length is of the order of nanoseconds and the current amplitude a few amperes, enough to destroy susceptible microcircuits.

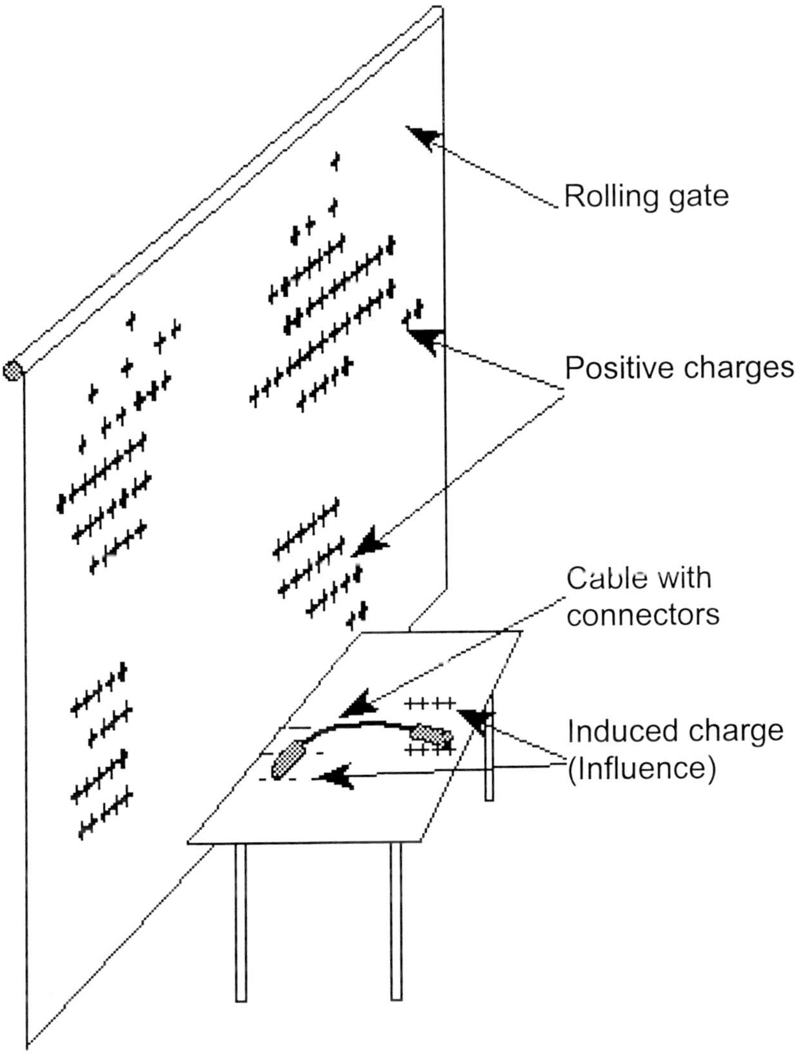

Fig. 35. Rolling gate as charge generator. ESD-hazards are highest at new gates. In time the stuff will attract dust and dirt and absorb water. Then dissipation of charges can occur, lowering the voltage

2.6 Relation between charging - discharging processes and ESD-simulation models

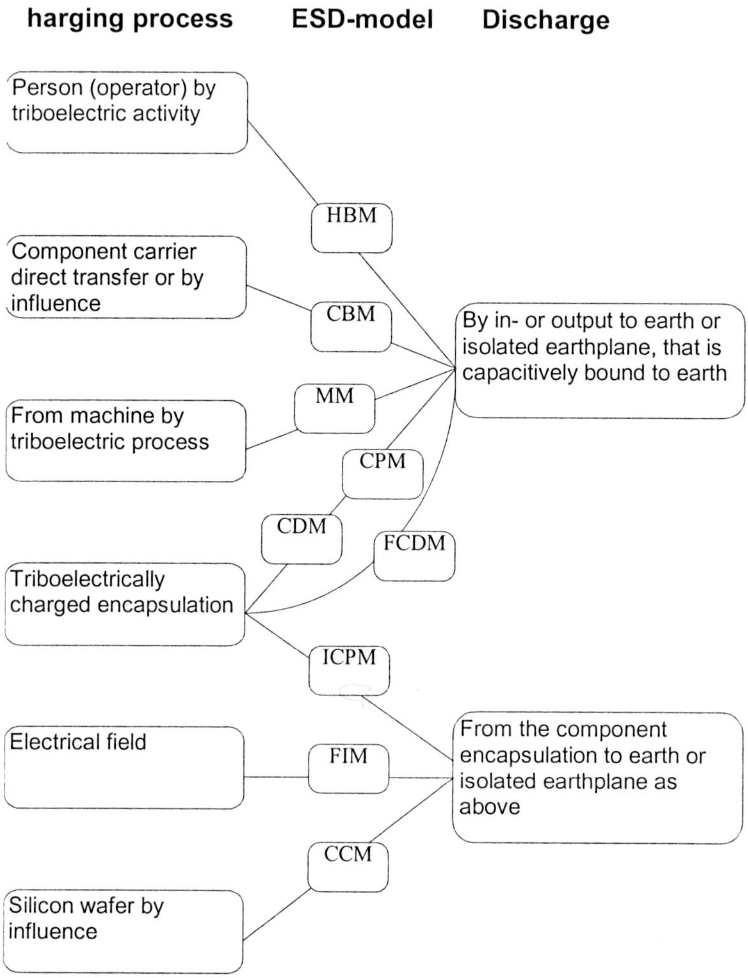

Fig. 36. Connection between charging generators and ESD-models, described in next chapter

3 ESD-testing and models

3.1 General

In the 60-ties a very high vulnerability of the new discrete MOS-components was realised. They failed often without any apparent reason. A thorough failure analysis of damaged components started to settle failure mechanisms. This led to the understanding that electrostatic discharges caused voltage breakdown in the thin silicon gate oxide of the MOS-circuits. The accidents happened during the handling of the circuits. The operators had been charged triboelectrically and by touching some pin of the circuit a discharge destroyed it. The thickness of the oxides was normally about 1000 Å, corresponding to a breakdown voltage 100V. The failure mode was often a short circuit between gate and source or drain. About 50% of all failures were caused by ESD.

The failure mechanism clarified preventive measures had to be taken. All personnel handling the circuits in production, control and packing at the manufacturers, respectively unpacking, testing and mounting for use in equipment fabrication by the receiver could harm the circuits. They get charged in this normal work and cause discharges when handling the circuits. The results can be, if not catastrophical, that some parameters are changed making the circuit unusable for intended application.

The first action was to protect the MOS-components by keeping them in a metal foil until they were to be mounted. This implied some protection as the components were shielded against external fields and discharges to the gate oxide. But hazards remained still at wrapping and unpacking the items, testing and mounting them.

The problem was attacked in different ways. Protection circuits could be designed and implemented in the component wafer. The first protection circuits had resistors and capacitors as elements. Experiments were also performed with spark gaps. The function of the protection circuits is twofold, to reduce the pulse height to a non-dangerous level before the pulse reaches the gate and to dissipate the pulse energy in their resistors.

Many investigations started in USA to find out under what circumstances people get charged and to which level depending on different activities. At the same time a more effective design of protection circuits began. They became faster, withstood more pulse energy together with less space on the wafer. The inputs of the circuit being most sensitive first got protection.

Methods to test the susceptibility of components for ESD came to be considered, including the efficiency of the protection circuits. Modern microcircuits can sometimes be protected against several thousands volts. This has been realised for most CMOS-circuits.

Methods for failure analysis of components damaged by ESD were developed. They aimed at localising the failure site and clarify failure mechanism. Being regarded as a big problem for the electronic industry some kind of cooperation to study the ESD-phenomena felt necessary and a symposium on the matter was initiated in 1979, the first EOS/ESD-symposium in USA. These could be repeated annually and formed a forum for treating ESD-problems. Around 1500 participants from both USA and Europe and Japan confirms the importance found for dealing with ESD-phenomena. In 1982 the US EOS/ESD-association was established cosponsoring the ESD-symposia. EOS means electrostatic overstress and represent the conventional transients.

3.2 ESD-simulation models

3.2.1 Human Body Model, HBM

The discharge from operators to electronic circuits described above became the background for the first ESD-testing model. On this model simulators were constructed to measure the ESD-susceptibility of electronic components. The simulator produces electrical pulses of the same shape as these from natural discharges. The human body is represented by a capacitor and a serial resistor. Because of his capacitance to earth a person can, if insulated, keep this charge until a contact to earth is made. Metal parts have the function of "local" earth.

The human body is also able to conduct current, the outer parts of the skin is normally a current path and represents a resistor. The value of this resistor will of course vary from person to person. For the Human Body Model (HBM) it was decided to have a capacitance of 100 pF and resistor of 1500 ohm as standard values. The real capacitance for the human body is of course also no constant. Values as high as 500 pF and low as 70 pF have been measured. They are also dependent on the geometric relation body to earth. The HBM simulation, implying charging a capacitor and discharging it through a resistor to a component, is a simplification. Both capacitance and resistance are distributed in the body and some inductance action of the body occurs. The model should, to be more correct, include a complicated net of impedances, the magnitude varying from one person to another.

The HBM simulator has, however, remained the same for about twenty years. It has become standardized and is described in MIL-STD-883 C method 3015.7, "Test Methods and Procedures for Microelectronics", and the US ESD Ass. standard for HBM.

Main characteristics are for the realised pulse a rise-time of less than 10 ns and pulse length of 150 ± 20 ns. This wave form shall be checked using a current probe and the component socket short circuited. The effective capacitance is to be measured by an electrometer. In "Notice 7" is also stipulated a maximum value for so called ringing of the pulse, 15% of the amplitude.

Failure criteria for components tested by the HBM-method can be:

a) any parameter gets its value moved outside the specification range
b) any parameter value is changed 10% or more after the test

This means that a component whose leaking current exceeds the maximum allowed value after test is disqualified. If in a test a transistor current magnification factor changes 10%, the transistor will be disapproved, though, maybe, the factor is still within the specified range.

Different HBM-simulators can give different test results even if they are of the same manufacturer. This can be due to only small differences in rise-time or pulse energy, being the crucial conditions for a damage to arrive or not.

A typical case for HBM is given in fig. 37. Standard HBM circuit for testing is found in fig. 38 and standard pulse shape in fig. 39.

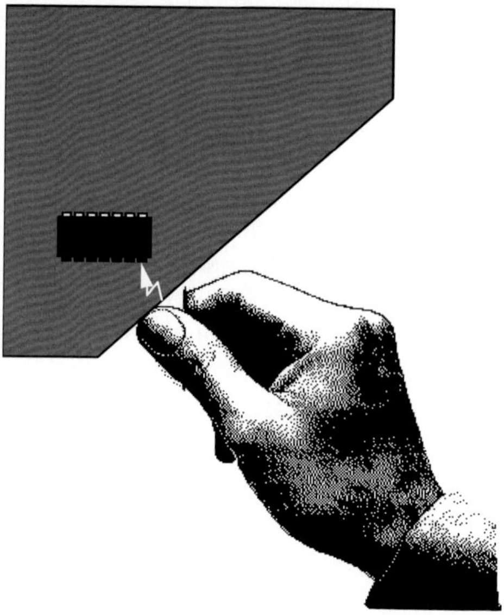

Fig. 37. The picture shows the kind of discharge simulated by HBM: discharge from a person to a component

The chosen values of resistor and capacitor correspond, as mentioned not always to empirical discharge conditions. Experience seems to show that the resistor value is too high. An absolute value for the ESD sensitivity for HBM-discharge is thus difficult to obtain and the method is therefore "mostly used as a reference to compare the sensitivity for different components".

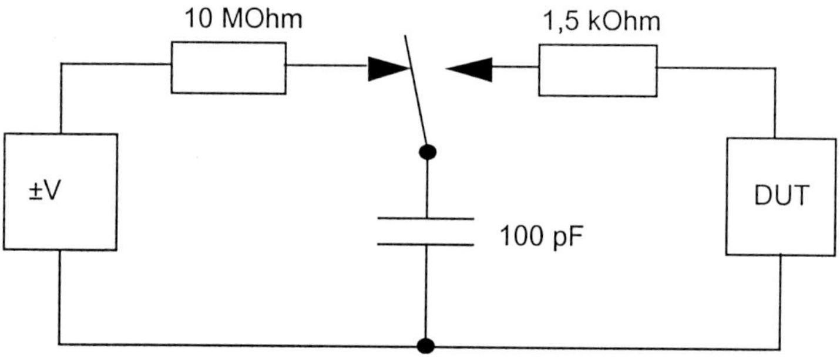

Fig. 38. An outline of the simple HBM-scheme. The capacitor can be charged positively or negatively from the high voltage unit (HV). When switching the relay the capacitor is discharged through the resistor and DUT (Device Under Test). The resistor will limit the current. The charging of the capacitor takes some milliseconds (ms) while the discharge has passed within about 200 nanoseconds (ns). This makes the whole construction critical. Rise-time cannot exceed 10 ns. The length of the conductors must then be as short as possible to limit their inductances, influencing the rise-time. Another critical factor is the dependence of the capacitance on the voltage. Many capacitors will loose up to 40% of their capacitance when connected to a high voltage. Thus less energy will be transferred to the DUT

In many commercial HBM ESD-testers the influence of the contribution to the capacitance from the relays has not been sufficiently analyzed. The effect can be 15-20% too high capacitance.

Inductances of the conductors and capacitive interaction between conductors and relay contacts will often affect energy and the discharge time for automatic ESD-testers. According to MIL-STD-883 D, method 3015.7 (March 1989), input and output of an IC-component shall be tested against all others with five positive and five negative pulses. One second delay between the pulses will give sufficient cooling. Compare also with the ANSI EOS/ESD S 5.1 - 1993 standard. All this testing will take at least one hour or in reality about one working day to manually investigate a 20 pins DIL! Fortunately automatic HBM testers were elaborated in the last 10 years making the measuring time much shorter.

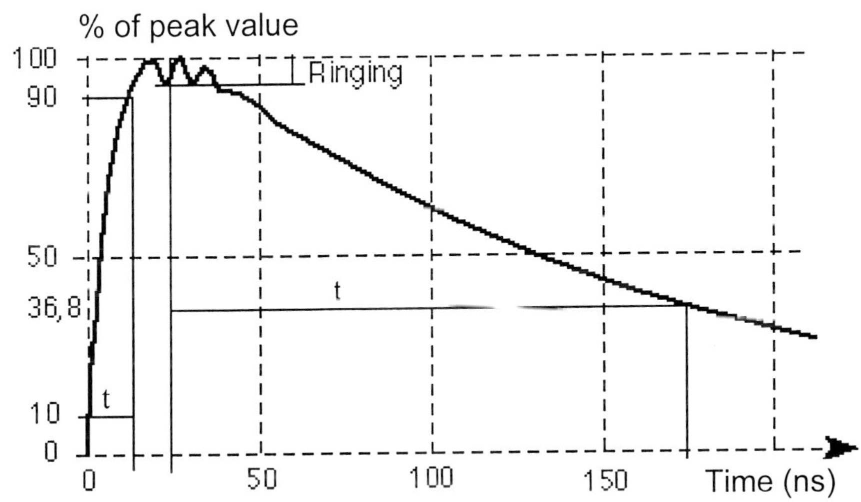

Fig. 39. HBM voltage pulse according to MIL-STD-883 D the pulse shape can be described by:

$$V = V_o \cdot e^{-\frac{t}{\sqrt{LC}}}\left(1 - e^{-\frac{t}{RC}}\right)$$

t = rise-time (≤ 10 ns, measured between 10% and 90% of peak value)
t = pulse length, measured between peak value and 1/e of this value (C x R = 150 ns). Ringing at most 15% of peak value

3.2.2
Machine Model MM

When components started to be manipulated in machines - testing equipments at arrival inspection, automatic mounting robotics in production and automatic PCB-testers by final inspection - an increase in damages was notified. Failure analysis afterwards revealed that ESD often was the reason. Many microcircuits that were destroyed had passed the ESD-test according to HBM method! Yet they collapsed. The input protection current remained in general intact, though the main circuitry got harmed. This indicated a new kind of discharges compared with HBM. Investigation performed in Japan proved that the component handling equipment generated these very hazardous discharges.

Parts of the testing or measuring equipments became evidently charged and when pins (connectors) of components got in contact with them a discharge was released hurting the components. From a number of measurements a capacitance value of the magnitude 200 pF was found for the charged equipment parts. The discharge resistances, often metal to metal, proved to be very low, in reality that low that no extra resistor is necessary in a simulation model.

These investigations resulted in a model with a discharge circuit consisting of a 200 pF capacitor + "zero ohm" discharge resistor, named the Machine Model, MM. The discharge pulse are much faster compared with those of HBM-method. The latter gives a pulse time constant of about 0.1 μs, while that of MM is of the order some tens of nanoseconds (ns). The fast discharge makes that the protection circuits, effective against HBM pulse, not have time to react and other protection means must be used.

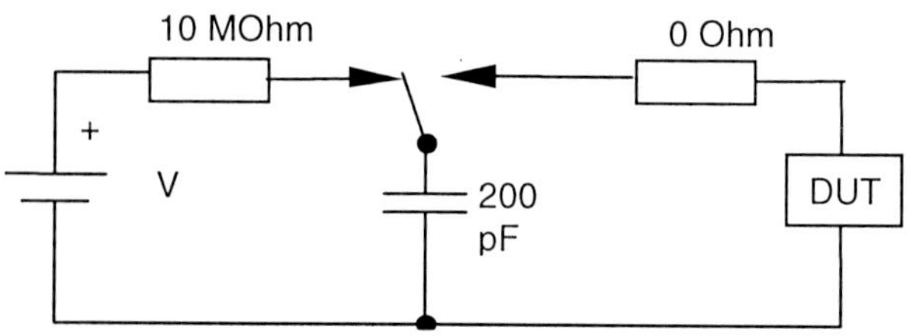

Fig. 40. Schematic outline of MM

The Machine Model has the same configuration as the Human Body Model, only the capacity is 200 pF instead of 100 pF and a serial resistance as low as possible. This model can be considered as a "worst case" of HBM.

In the machine model simulators it is difficult to achieve sufficient low inductances, especially in automatic ESD-testers of this design. Thus the spread in results from different testers is large. Comparison between tests with MM and

HBM show that components that are strong against HBM, "often are very sensitive against MM". Standard pulse shape for MM is given in fig. 41.

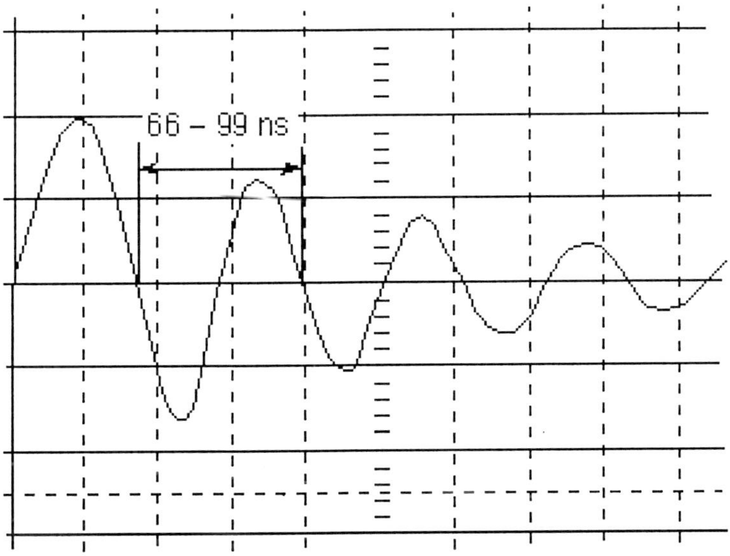

Fig. 41. Discharge pulse from MM-tester. According to US EOS/EDS Ass. std DS5.2, the time elapsed between the first and third "zero passage" shall be 66-99 ns.

This model is applied rather much in Japan and Europe, as "second ESD model", while in USA the following model, CDM, is preferred after HBM.

3.2.3 Charged Device Model

Encapsulated components, microcircuits, will be triboelectrically charged by friction or separation from other insulators. As described before metal parts will undergo charge separation by influence from the charge on their encapsulation. If some of the pins is brought in contact with a metal piece or earthed otherwise a discharge will take place. A model for such an event was presented first time in 1974 (Speakman 1974), the Charged Device Model. The rise-time of the discharge pulse in this case is much shorter than at HBM. Input protection circuits are not effective against these discharges. They are too slow and discharge pathes are different from the HBM case. Interior parts of the circuits can be damaged. Fig. 42.

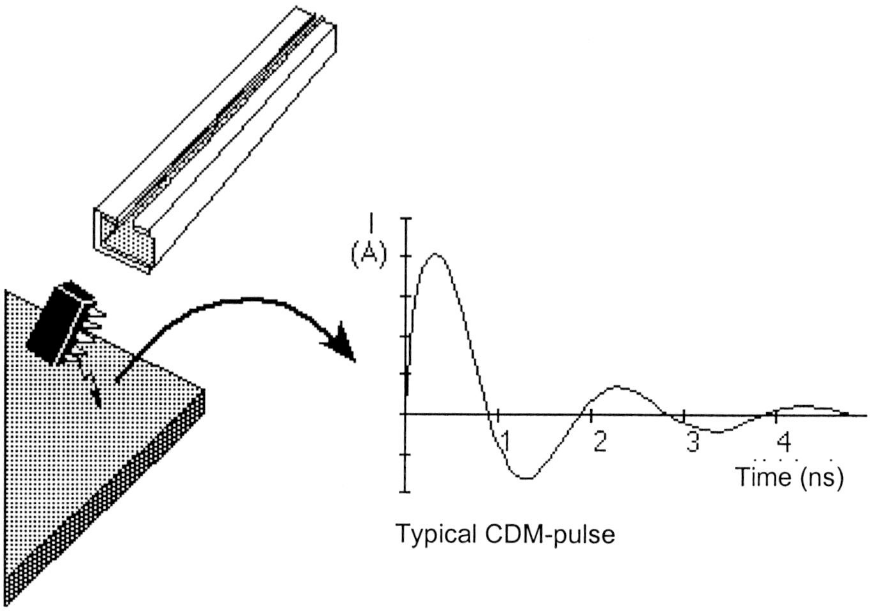

Fig. 42. The picture shows an example where simulation with the CDM-method is to be used. Because of sliding friction and separation in the component tube, the encapsulation and the pins get charged. By influence from these charges conducting parts will undergo charge separation. If a pin - usually a "corner pin" - comes in contact with bench surface (or some other conducting object) the component is rapidly discharged. Currents of magnitude 15 A for about 10 ps are released (1 ps = 10^{-12} sec). The voltage can be high enough for oxide breakdown and the energy sufficient to destroy an pn-junction of a bipolar transistor.

An encapsulation with "floating voltage" metal lid are often sensitive to CDM-discharge. Experiments with 16 pins DIL, charged with 2 nC and discharged through an "one ohm" resistor showed a rise-time of 0.5 ns and pulse-length of several ns, peak current 15 A giving a very high power density, about 10^9 W/cm^2. (Bossard, Chemelli and Unger, 1980). Much earlier a formula was presented, that describes what value the power density must reach to damage a component by melting. Pulse length, melting temperature, initial temperature, heat-conduction ability, material density, current distribution and voltage drop etc are parameters composing this formula. See fig. 43.

Automatic CDM-testers have been developed. Some have several options: CDM, HBM and also latch-up testability for instance. Charged device problems have been the object for many studies.

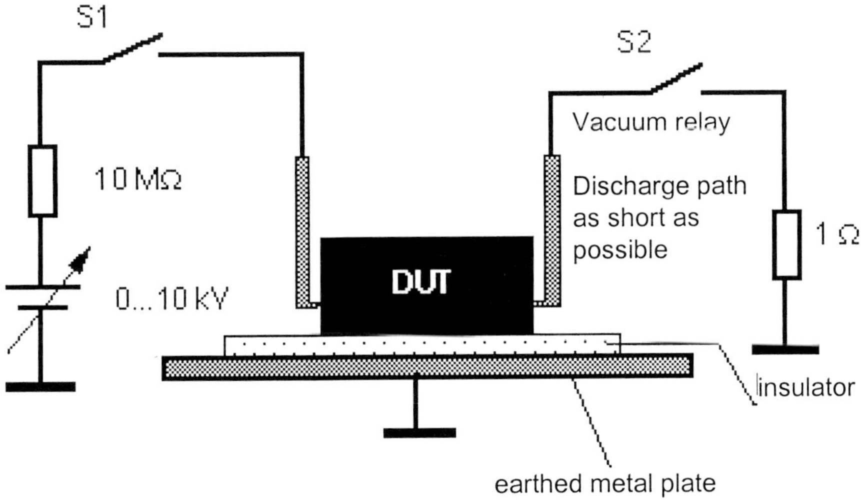

Fig. 43. Charged Device Model
Testing sequence
S1 and S2 open at the start
S1 closed, S2 open at charging process
S1 open, S2 closed at discharging process

The presence of inductances in the ESD-tester is critical. A chip resistor, for instance, includes 1 nH inductance. Only about 2 nH will erase the difference between tested SOT- and DIP-components.

Induction (influence) can also come from test sockets, that consequently should be made by antistatic material with high resistivity, eliminating short circuits.

Testing standards for different test modes (HBM etc) are available from the US ESD-Association.

In fig. 44 the shape of a theoretical decay curve is given.

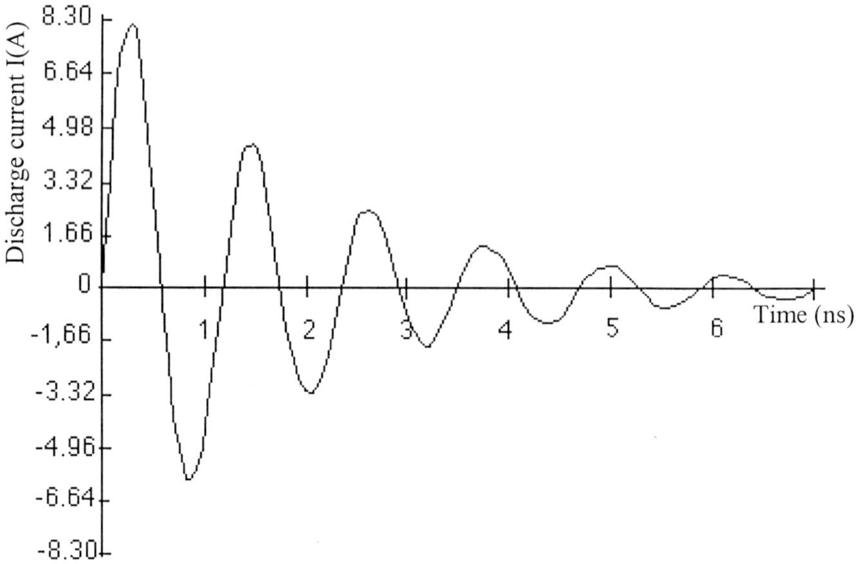

Fig. 44. Calculated current as a function of time:

$$I(t) = \frac{V}{\varpi L} e^{-at} \sin \varpi t; a = \frac{R}{2L}; \varpi = 2\pi f$$

Parameters chosen: R = 10 Ω; L = 10 nH; C = 3,6 pF; V = 400V
C is the capacitance of the component package to earth and determines the potential of it at a certain charge.

3.2.4 Charged Package Model, CPM

A sophisticated variant of CDM is the CPM, Charged Package Model, that was elaborated to eliminate the effect of humidity as one reason. The time elapsed between charging and discharging of a component will affect the result at CDM, if the humidity is high. At CPM tests all pins are charged simultaneous through a high recistivity material, bakelite. See fig. 45.

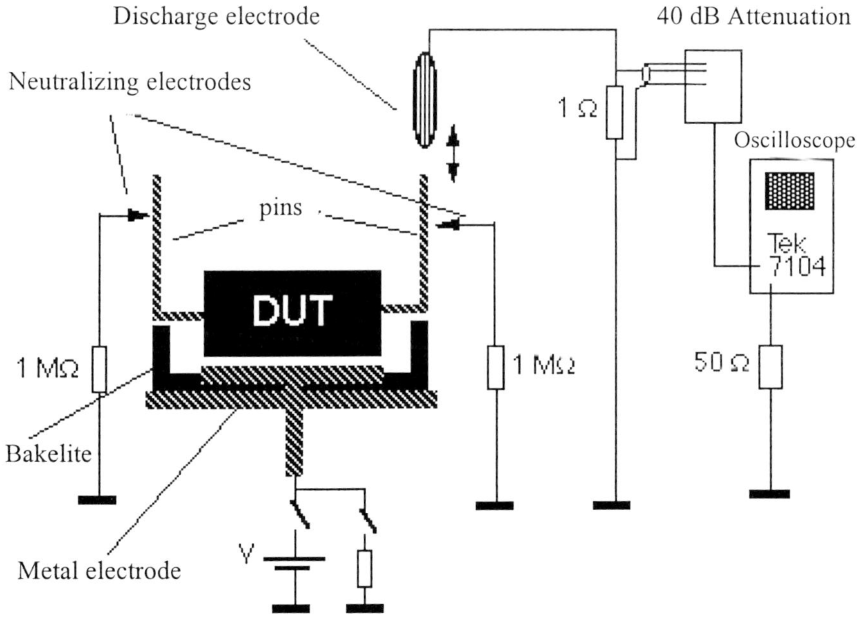

Fig. 45. Equivalent scheme of CPM

A special discharge electrode is lowered to the connector (pin) to be tested. Discharge is accomplished through an 1 ohm resistor to earth. The pulse shape can be studied by an oscilloscope. Before next charging of the component all connectors are neutralized by a two file set of electrodes. Thus all possible remaining charge is led to earth by an 1 MΩ resistor. The test set-up is then ready for a new procedure - charging, discharging, neutralization etc. In described CDM tests the pins are charged by connection to a high voltage source. But there are other approaches, trying to reproduce the true charged device process

3.2.5 Charged Board Model, CBM

There was for a long time misunderstanding about components mounted on PCB:s and ESD. They were thought to be protected against ESD! As a matter of fact they can be damaged at lower discharge voltage compared with the CDM- or HBM-testing of the single component! Crucial is the capacitance between the PCB and earth. (Enoch and Shaw, 1986). See fig. 46.

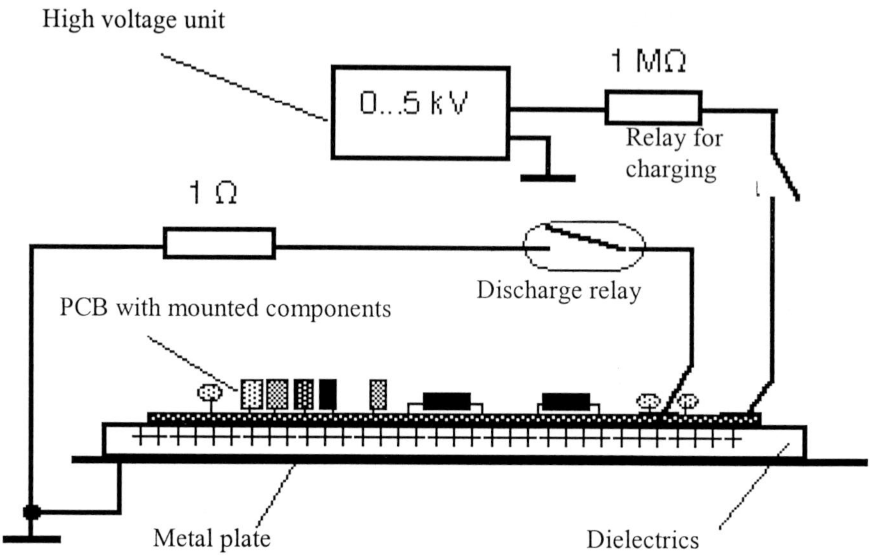

Fig. 46. The picture illustrates CBM, Charged Board Model. The PCB to be tested is placed upon an insulator plate (dielectrics) with an earthed metal plate as support. The PCB is charged from the HV-unit and discharged via a connector and 1 ohm resistor to earth

There are two very distinguishable cases for the occurance of failure on a PCB. The first one can be described by a charged person who is touching with his hands a board, direct or capacitively earthed. The second occurs, when a person is carrying a board with himself and gets charged. The discharge will come afterwards, for instance on placing the board into a stand. Part of the charge on the board will discharge rapidly when brought in contact with metal parts of the magazine.

The reason for discharges of PCB:s is the formation of charges on metal parts (conductors) by induction (influence) from the charged board, usually a dielectric.

3.2.6 Field Induced Model

The possibility to induce charges on a component by an electrical field was treated before. The field can have its origin in charges on plastic materials, as coffee cups, packaging materials etc. If these objects come near electric circuits charges will be induced on the circuits damaging them if the charges can recombine by some short circuit. A model, FIM, for these process is given in next figures.

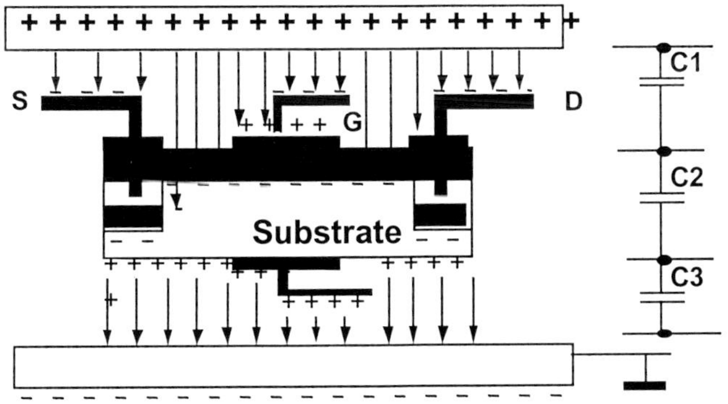

Fig. 47. This picture shows the idea of the Field Induced Model. A component e.g. a microcircuit, inside an electrical field will submit a charge separation like other objects in the field. If one side of the charged circuit gets a short circuit to earth a discharge current starts flowing that may cause current induced defects

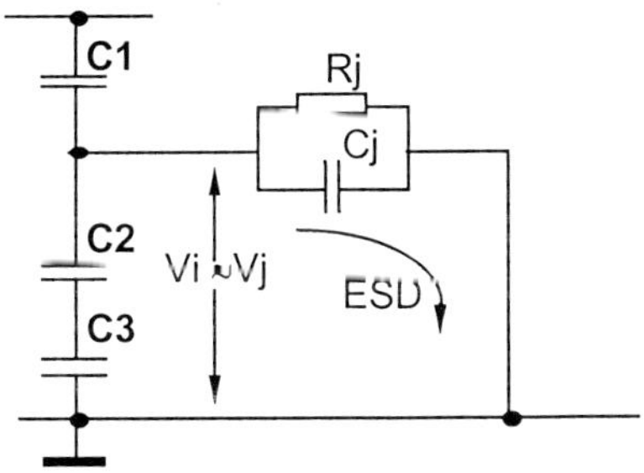

Fig. 48. Equivalent circuit scheme for FIM. R_j and C_j represents the DUT.

3.2.7 Charged Chip Model, CCM

Charged Chip Model is used to test the ESD-sensitivity of silicon chips. The chip is charged capacitively and discharged afterwards by an electrode approaching the chip until contact is established. This method simulates the picking up of semiconductor chips with vacuum pliers. See fig. 49.

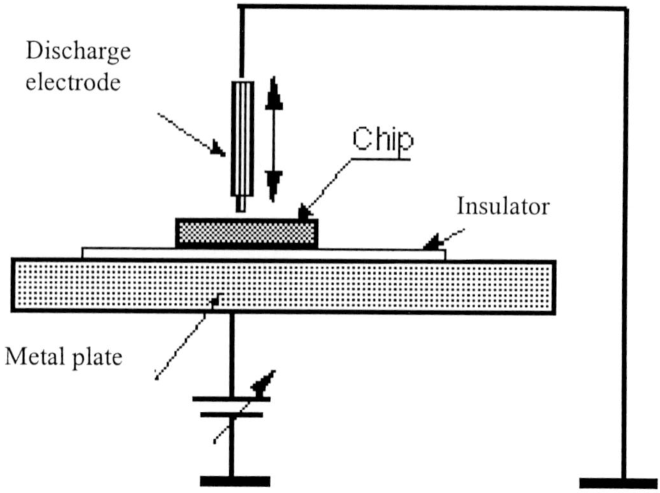

Fig. 49. Principle skiss of Charged Chip Model

3.3 Comparison between different models and the components' responses to them

Usually components are tested according to the HBM-method. Though the CDM-method is gaining more and more interest. The latter has a faster rise-time and are more severe for the components than HBM. The discharge pathes are different in the two cases. Then circuits should be protected against both of them.

It is not possible to get some interelation between results of HBM- and CDM testing. But for different values of capacitance, C, and resistance, R, in performance of the HBM method a relation was derived (VZAP-1, RAC/RADC, 1984):

$$V_2 = V_1\sqrt{\frac{R_2 \cdot C_1}{R_1 \cdot C_2}}$$

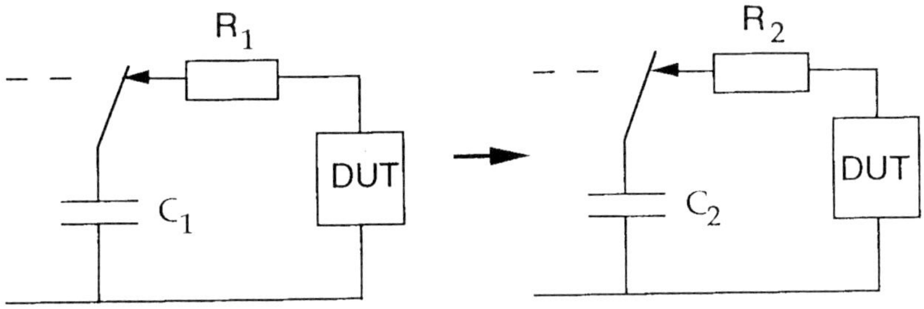

Where V_1 is the threshold value of damage in the first combination and V_2 in the second. According to VZAP-1 is

$$\sqrt{\frac{R_1}{C_1}} = 3{,}87$$

These results are empirical but in agreement with an analytic approach based on an energy model. This uses the assumption that the damage is caused only by the energy transferred. The relation was proved to hold for results with the standardized HBM (C=100 pF, R=1,5 KΩ) and the earlier IEC-model (C=150 pF, R=150 Ω)

As the MM-method is also used, significant results, of MM and HBM are demonstrated in next figure.

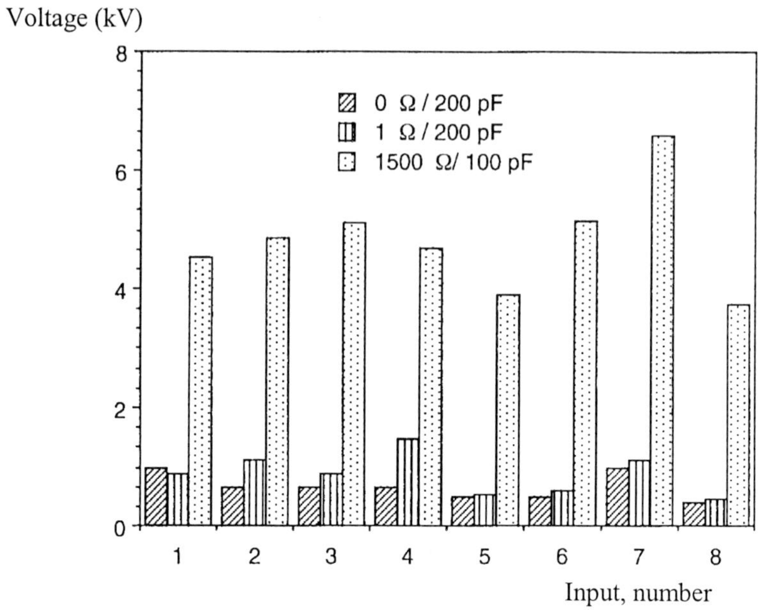

Fig. 50. The diagram shows results of experiments performed with HBM and MM. In the latter case both a zero resistor and 1Ω resistor was used. An ASIC CMOS especially designed for ESD-tests was measured. The susceptibility level for HBM and MM are very different, 3500-7000 V for HBM, while less than 1000V for MM

Further investigations proved that the failure mechanisms for HBM and MM are the same. Then these models can be compared. But CDM is very different from them. It must also be emphasized that a comparison between HBM- and the faster MM-method should be done with caution. This depends partly on the fact that exposed components can have parasitic capacitances and restistances of the same order as the testing circuit! Should the testing circuit have an initial high transient , the result will be erroneous.

Results how the ESD-susceptibility can depend both on the parameters of the testing circuit and design parameters of the component, as doping depth, were presented (Maeda and Wada, 1989).

4 Susceptibility for ESD

In respect of the performance and included application of electronic components different requirements as to ESD-susceptibility are given. Some are given in MIL-Std-883C others in STACK, a mainly European norm. Test performances are found in them. This cannot prevent any company or other users to formulate its own requirements. Fig. 51 can serve as a guidance. Fig. 52 gives thresholds.

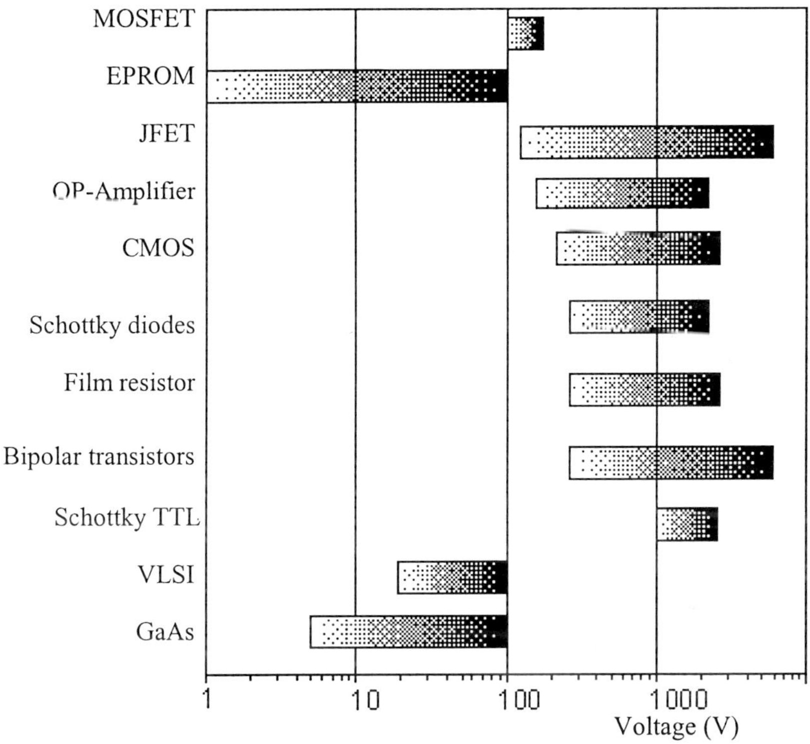

Fig. 51. ESD/HBM voltage susceptibility of different technologies the range to the left of the light parts is not risky for the component while that to the right of the dark range implies high vulnerability the shadowed parts correspond to susceptibility variation of protected circuits

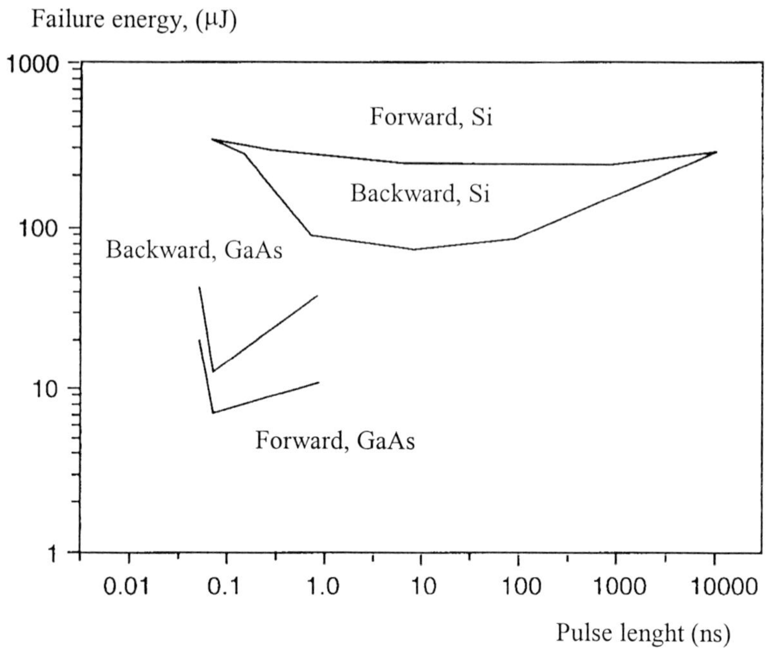

Fig. 52. Failure threshold energy for silicon, respectively, Gallium Arsenide components, backward and forward current direction.

The specified ESD-requirements are based on the HBM-ESD-susceptibility and represents discharges from an operator to the component. Extensive determinations of ESD-susceptibilities for components were performed at the RAC/RADC-reliability center in Rome, NY, USA. They are published (see p. 168).

The diagram indicates an extreme ESD-sensitivity for GaAs-components. Many optocomponents are made of GaAs (lasers for instance). This is due to material properties, see figures 53 and 54. Heat conduction decreases rapidly with increasing temperature for all semi-conductor materials given in figure 53. The higher the temperature in the material the less the possibility of heat transport.

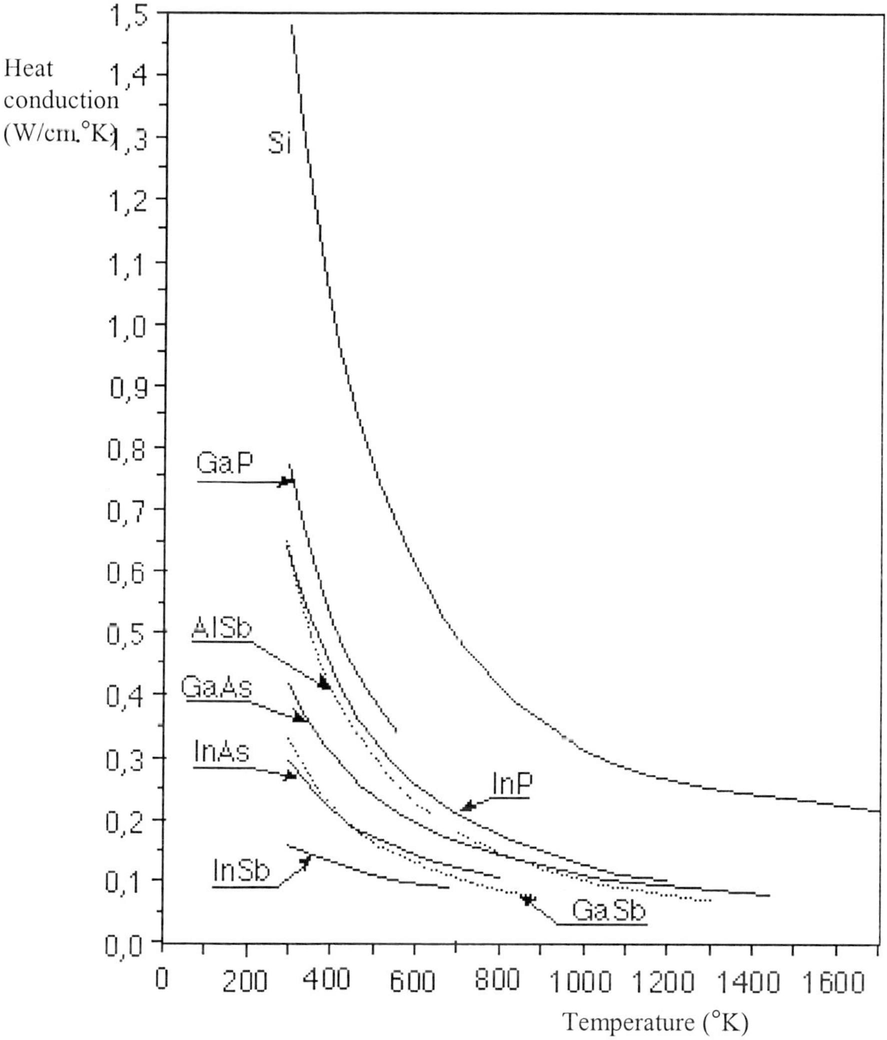

Fig. 53. Calculated heat-conduction as a function of temperature for different semiconductor materials

The larger density a GaAs compensates partly for the larger heat-capacity of Si, but the latter has a higher melting point.

Though GaAs-circuits are regarded as high temperature components they endure heat less and are therefore much more vulnerable for ESD than silicon based circuits. This is valid for most III-V materials. The diagram in figure 54 gives the melting temperatures for semiconductor materials. That of Indium-Antimonide (InSb) is only 530°C!

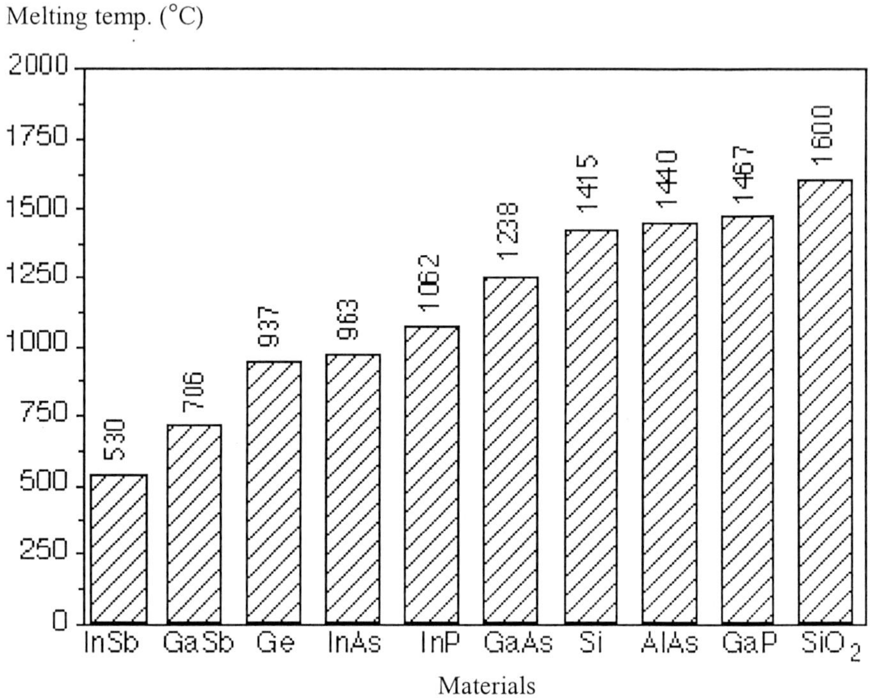

Fig. 54. The effect of an ESD-pulse, the energy of which is developed in the bulk of the material, depends on material properties and dimensions, i.e. if a melt damage will occur or not. The crucial question is, if the melting temperature will be reached or not. By sufficient good heat conduction the heat energy is carried off. Thus the heat conduction ability is important because ESD-pulses can cause a rapid temperature rise

ESD pulses propagating through semiconductor surface conductors can give rise to melt damages, the current being sufficient high. Other damages are also possible. When melting the conductor material (usually aluminium) will dissolve in the material under the conductor. At a certain concentration of the conductor metal a so called eutecticum is formed, having a lower melting point than each of the two materials. A melt of aluminium and silicon is created diffusing into the silicon. Reaches the melt an pn-transition, this will get a short circuit and the damage becomes permanent. See fig. 55 and 56.

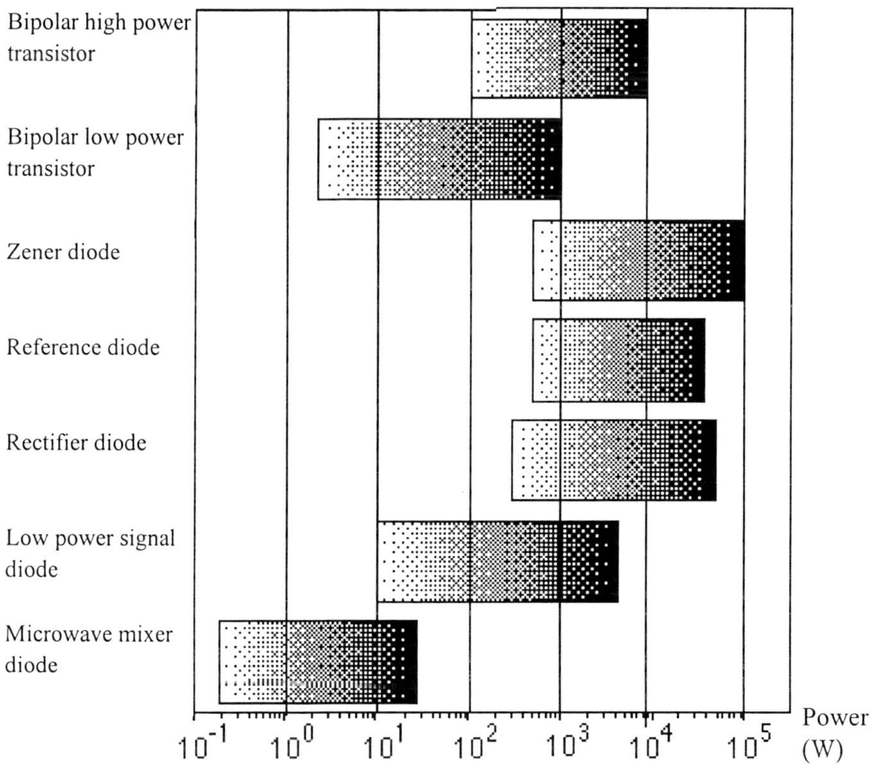

Fig. 55. The diagram describes the variation in susceptibility of different semiconductors for µs transient pulses of different energies. The energy range for damage is given

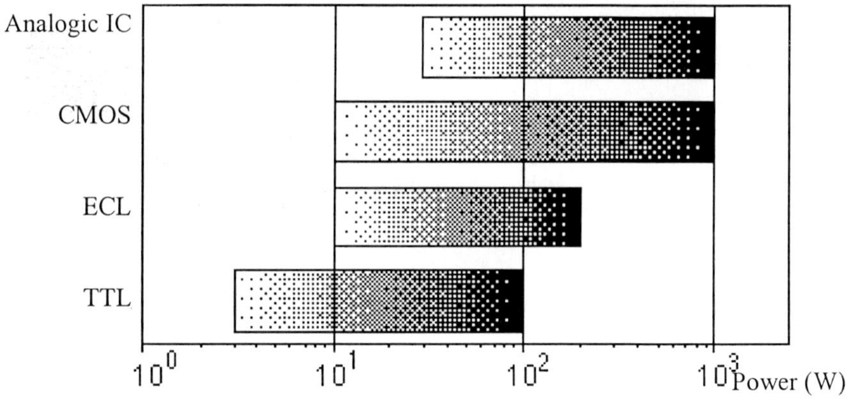

Fig. 56. The sensitivity for overloads of ESD-type for a selected choice of integrated circuits of different technologies. The diagram shows the variance in sensitivity for transients of µs pulse length and different power

Aluminium is the dominating conductor metal for silicon based semiconductors. For Gallium Arsenide components a multi-metal system is often used, as gold/chrome with gold/germanium/nickel as contact material.

4.1 Future implications of ESD

Two factors indicates that ESD will be a common cause for failures in electronic components.

> ESD-discharges of type HBM will be less important as efficient protection circuits are easy to achieve. But protection against discharges of type CDM are more difficult to obtain and they will be more common as automatic handling will increase of cost reasons.
>
> Discharge of type CDM are very fast with a rapid temperature-rise in the semiconductor material. Modern microcircuits will have close-packed structures and material with worse thermal properties, improving the hazard of failure very much. The CDM discharge procedure depends very much of the encapsulation, too.

The trend in the development of electronic components is towards closely packed structures, more elements per surface unit and use of thinner gate oxides and new materials with lower heat conduction properties, e.g. gallium arsenide. All this will increase the ESD-susceptibility. The following table gives an illustration about packing density development for memories.

Year	Laboratory	Product introduction	Full production
1986		4 Mbit/0.8 µm	1 Mbit/0.8 µm
1990	64 Mbit/0.35 µm	16 Mbit/0.6 µm	16 Mbit/0.6 µm
1997	256 Mbit/0.18 µm	64 Mbit/0.28 µm	16 Mbit/0.4 µm

The narrow submicom conductors are realised in multi-metal layers. The increased automization in mounting and testing of components was already mentioned. An implication of that will be high demands on testing systems and equipments, as insulating parts easily get charged to high voltages, causing discharges. Metal parts have low resistances resulting in high discharge currents, increasing the risk for ESD-damages.

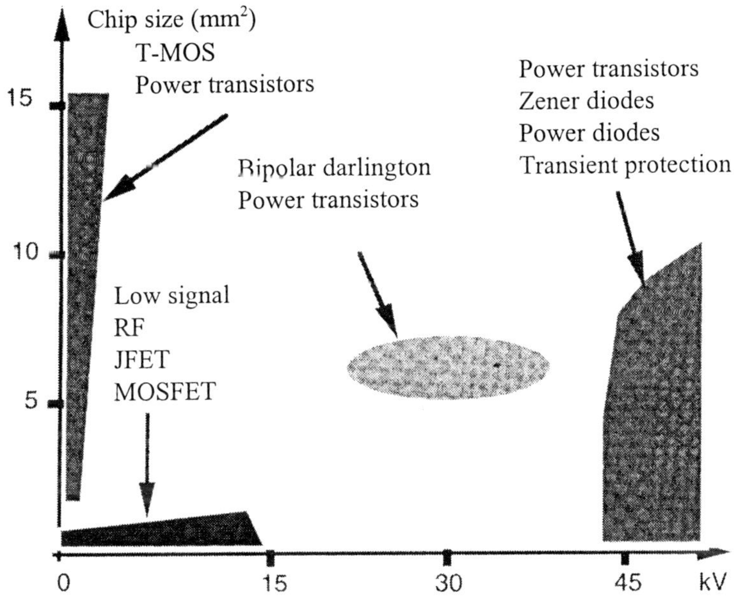

Fig. 57. Voltage solidity for some different semi-conductors as function of chip size, according to HBM-tests

How fast the circuits can react depends ultimately on the mobility of the charge carriers in the semi-conductor material. That is why the interest is focusing on III-V materials with high mobility carriers as gallium-arsenide and germanium-silicon. The drawback with these materials was already pointed out, high thermal resistance, lower specific heat than silicon and often a lower melting temperature.

Altogether the ESD-susceptibility might increase dramatically for components made of these materials.

GaAs MESFET with gate length 1 μm, for instance, measured according to HBM failed at about 200 V. Old MESFET could fail already at 50 V, making handling of them impossible without effective ESD-protection. Other sensitive opto electronic components are lasers and micro-wave components as wave-guide modulators.

A theoretical study, based on the Wunsch-Bell and Siemens method, about the expected ESD-susceptibility for components of III-V-material has been performed (Chase, 1988). The result could be interpreted as components of III-V materials can have about ten times higher ESD-susceptibility than silicon components.

4.2 Example of determination of susceptibility level

With susceptibility level is meant the voltage pulse level, at which a component becomes damaged. If a certain voltage is chosen the result need not necessary be a damage. There are schemes given in test standards for the choice of start test value and increments for repeated stress until a failure occurs.

An investigation of tests with HBM in 150 pF, 150 Ω combination will be referred to. The aim was to find out, if there is a relation between voltage endurance and input capacitance of HEXFET components. See fig. 58.

It seems natural with such an approach, to find relations between transient solidity and input capacitances, partly as these capacitances vary very much, partly because the capacitive property implies collection of charges.

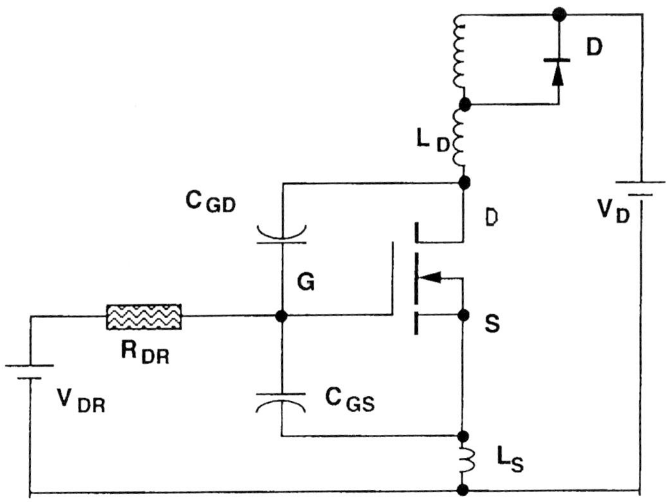

Fig. 58. Scheme of a typical HEXFET input. Input capacitance $C_{ISS} = C_{GS} + C_{GD}$

Seven groups, each with 10 components, got their input capacitance C_{ISS} measured. Afterwards they were exposed to transients until parameters changed to be outside the specification. After mean value calculation the values were plotted in a diagram, se next figure 59.

The results are represented by a remarkably straight line. Regression analysis made with the simple linear approach (Index ZAP refers to the parameter of the HBM circuit.):

$$V_{ZAP} = A + B \cdot C_{ISS}$$

gives the approximation:

$$V_{ZAP} = 91.8 + 0.175 C_{ISS}$$

A polynom of 4th degree will almost completely fit to the experimental values. That is mathematically beautiful but will not explain the physical relations.

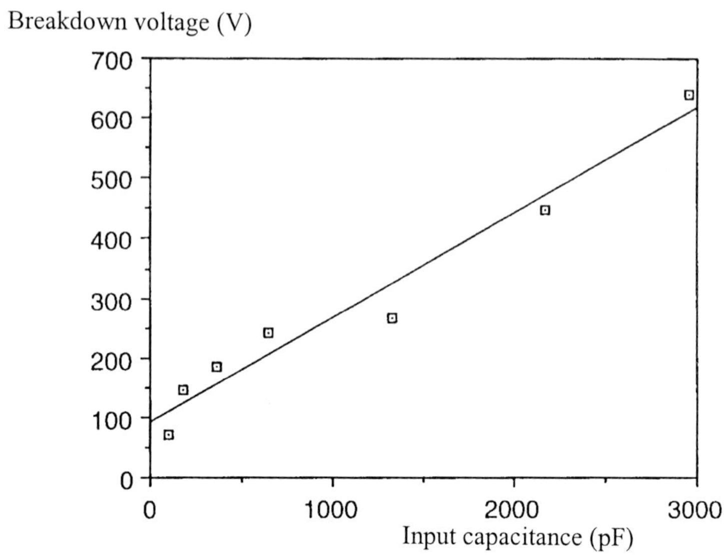

Fig. 59. Breakdown voltage as a function of different HEXFET input capacitance.

The following partition of a capacitor charge is assumed to be valid:

$$Q = C_1 U_1 + C_2 U_2$$

the result will be:

$$U_{ISS} = V_{ZAP} \cdot \frac{C_{ZAP}}{C_{ZAP} + C_{ISS}}$$

According to preceding formula the U_{ISS}-value will increase with increasing input capacitances. There is a relation between U_{ISS} and the input capacitance.

A partition of the energy between the capacitive part of the input and the discharge capacitor will give the relation:

$$U_{ISS} = V_{ZAP} \sqrt{\frac{C_{ZAP}}{C_{ZAP} + C_{ISS}}}$$

The result is described in figure 53. As U_{ISS} seems to be rather constant as a function of input capacitance the energy approach should be correct.

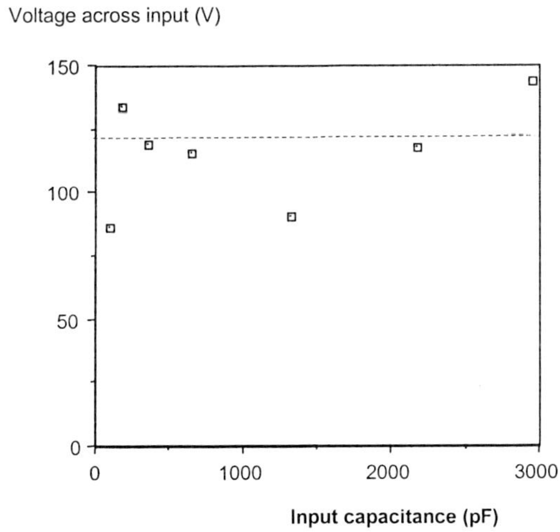

Fig. 60. The endurance against transient voltage across the input as a function of input capacitance.

This result can be used to estimate the susceptibility level, if the input capacitance is known. According to the diagram in the figure $U_{ISS} \approx 120$ V and from the formula for U_{ISS}, V_{ZAP} can be found:

$$V_{ZAP} = U_{ISS}\sqrt{\frac{C_{ZAP} + C_{ISS}}{C_{ZAP}}} = 120\sqrt{\frac{150 + C_{ISS}}{150}}$$

5 Failure analysis

5.1 General

By failure analysis it is possible to find out what has happened to a component that does not function properly and why. Only the failure mechanisms being known the cause of failure can be eliminated. Thus avoiding future problems. The more severe the consequences of a failure is the more important it is to deduce the causes.

The importance of failure analysis is elucidated by a comparison from another branch. The Englishmen were the first to use jet passenger aeroplanes, the British Comet planes. Unfortunately they became subject to a suite of disasters. Several of them exploded flying at high altitude. One above Himalaya, two over the Mediterranean Sea, for instance. One plane of the last two landed in shallow water and all essential parts could be saved.

It was stated that the plane had a breakdown before falling down. By an extensive investigation the failure mechanisms could be deduced. The plane had a window at the top of the cockpit. By this it was possible to navigate after the stars in case of urgency. It had sharp corners and was fastened in a frame by rivets. In such a corner a crack came up by fatigue. This crack grew at each change of pressure, that is at each starting and landing procedure. When it reached the row of rivets, when the strain in the metal was very high, it grew fast and "jumped" between the rivets. The metal weakened that much around the window that the whole window and part of the metal around broke away. As a consequence of the rapid pressure change the plane exploded and fell down.

This is a dramatic example of breakdown. Could something of that kind occur if only a transistor is destroyed? Probably we shall learn about still more dramatic accidents, or what is to say about a headline in the American magazine Electronics in the 60-ties: "60 cent IC near to cause nuclear weapon war". What worse could happen?!

The conclusion that a failing component from the field is a typical ESD damage can only be based on a thorough failure analysis. Time consuming work with effective analysis instruments is necessary. Earlier experience is a support to reconstruct the event. Data must be collected to find the factors influencing the breakdown, this to arrive at a certain conclusion. The more data found the more precise the analysis will be. The prerequisites for a correct analysis depends on the way of analysis - to do the right thing at the right time.

The analysis can often be time consuming and expensive, why it must start with a specification and goal formulation. A complete analysis of a complicated circuit can take several weeks to perform. If some data are missing, it will last much longer.

A simple strategy of failure analysis is used: All data (observations) from the component are recorded from outside towards the interior. For instance, it is not possible to measure the thickness of an oxide layer if it is already removed by etching!

A complete failure analysis is composed by the following five moments:

1. Specification of the analysis
2. Sign of failure at apparatus or system
3. Failure mode of component
4. Technical work with failure analysis
5. Determination of failure mechanisms and cause of failure.

5.2 Failure manifestation

If an effective checking system is applied a malfunction of an item gives rise to a failure report denoting the type of failure. If this indicates a separate component as the cause the separation is facilitated very much, but insufficient for a correct failure analysis.

Factors around the failure event are extremely important. It is desirable to decide at an early stage if the failure is of a systematic kind or accidental.

By "systematic" means that the failure occurred in a predictable way, as after long using time, elevated exploitation, bad component quality or systematical incorrect treatment.

During the life-time of an apparatus failures occur often accidentally in the beginning (early life). These can be defects from the fabrication or latent (see next chapter) ESD-damages etc.

Failures caused by ESD-pulses (transients) can followlingly be of both accidental and systematic character.

Important facts:
* Kind of failure, intermittent or persistent
* At intermittence, where and when
* Time of operation and environment
* The manufacturer of the component, denotation and date code
* Level of exploitation for the component
* Cause to remove the component

5.3 Failure mode

Distortion of the input characteristics of a semiconductor is often a sign of damage from some electrical overstress as ESD. Damages by ESD are different from other electrical overstress (EOS) damages. The physical impact is less extended but then also more difficult to localize. An ESD-damage is often revealed by an increased leakage current at the in- or output. The result can also be a decreased magnification factor at low current in a transistor or, simply a malfunction of an IC-component, leaving it impossible to localize the damage by measurements from the outside. ESD events can also result in melting or electromigration of , for instance, aluminium parts giving rise to disruption of conductors or short circuit for conductors to other parts of the circuitry or, finally, migration into an pn-junction. Breakdown of thin oxide layers is very common.

In operational amplifiers, usually some or both of the input transistors are affected and then commonly in the base-emitter junction. A consequence from this is often a too high offset voltage.

5.4 Technical failure analysis works

5.4.1 Preparation

To open microcircuits different methods are used dependent on the type of encapsulation. See fig. 61.

Ceramic encapsulations are often opened by simply introducing a sharp edge into the joint between the two halves and separate them. Round, metal encapsulation can usually be opened by a modified pipe cutter with a cutting trundle but they can also be opened by turning them in a lathe or grinding them.

Plastic encapsulation is often opened by a plasma etch or by a combination of grinding and dissolution of the plastic material. Suitable dissolving agents are heated, strong acids. Often sulphuric or red, hot fuming nitric acids are used.

> WARNING!
> All work with heated, strong acids and mixtures of different chemical agents can be very dangerous. Risk of explosions and corrosion wounds exists.

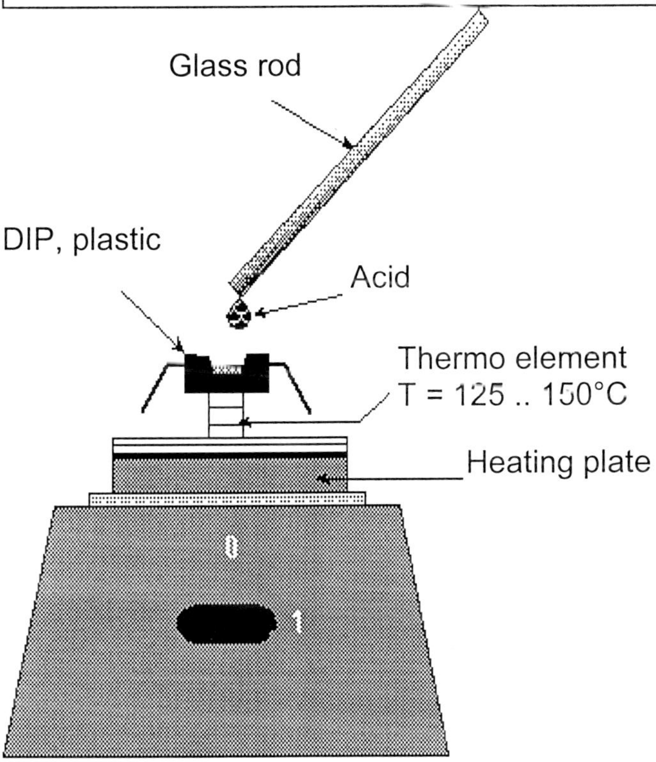

Fig. 61. The picture shows how a plastic encapsulated IC is opened by means of a heating plate and a metal plate with support for the encapsulation. Using a thermometer the temperature is regulated to about 125° and acid is dripped onto the upper side of the circuit just above the silicon chip. The acid dissolves the plastic encapsulation thus uncovering the silicon chip, that can be investigated.

The encapsulations once opened the circuits are examined, first in a stereo microscope with low magnification and afterwards in a microscope with rather high magnification and with enlightment of the beam path, a so called metal microscope. A scanning electron microscope (SEM) is used when there is a need for very good resolution. Also a TEM (Transmission Electron Microscope) is applied sometimes.

For study of, for instance, diffusion depths a polished intersection of the components must be investigated. Those are achieved by moulding the details to be studied in a suitable plastics and afterwards first grind them on wet grinding paper and then polish with, for instance, diamond paste on special disks for polishing purposes.

To study as an example pn-layers the surfaces are to be etched. This is performed using solutions of fluorine acid with a chemical buffer. The layers become visible as the solutions attack doped areas with different speed depending on doping concentration and doping elements. In other respects the technique is the same as at investigation of structures in metal alloys.

To find the failure site of a suspected ESD-damage in a component, it is often necessary to remove the passivation layer. This is often covering the main part of the silicon chip and the metal conductors on the surface. The passivation is often a glass layer possible to remove easily by some standard etch, based on ammonia fluoride, flourine-hydrogene acid and glycerine. Silicon nitride layers are often cumbersome to remove by etching. But one method is to apply hot phosphoric acid, another to use plasma etch.

Aluminium metalizations is often removed by a standard etch as for instance phosphoric acid, nitric acid and acetic acid in a water solution. It has to be observed that these solutions can be explosive!

5.4.2
Analysis instrumentation

Beside work with light-optical microscopes and SEM, normally used to localize damages in microcircuits, "hot spot detection" and "electron beam induced current (EBIC)" are utilized. Before was also TEM mentioned. Such one was applied in a recent investigation on damage from CDM discharge (Reiner, 1995).

By the electron beam in a SEM it is possible to produce pictures, based on the energy loss in a semiconductor material - a power development causing electron-hole creation. The size of the charge carrier current depends on the doping inside the material, while the thickness of the oxide layer and that of the passivation on the surface being parameters of the semiconductior technology. For this method there must be a good knowledge about the function of the component and the diffusion geometry. With this EBIC-method disruption of metalization, surface damages and diffusion defects can, as examples, be revealed. See fig. 62.

Fig. 62. EBIC-photo of an ECL-circuit showing leakage currents at outputs after ESD-stress of one input. It is not uncommon, as in this case, the other parts of a circuit get attacked, the part behind the stressed pin left unchanged

Methods for detection of so called hot spots (from increased currents) are *infrared micro spectroscopy* and *liquid crystals*. Another method was earlier used very much. It is based on local liquid boiling. Detected by this are bubbles formed in the liquid at the hottest surface spot of the object under investigation. Changing the pressure of the testing chamber, the boiling point can be altered, causing boiling already at a minor rise of the object's temperature. See fig. 63.

The method with liquid crystals is based on the transition between neumatic and isotropic phase in a thin sheath of liquid crystals. The transition appears in a rather narrow temperature interval. By this method it is then possible to detect small temperature changes at the surface of the object. In neumatic phase the liquid is transparent, getting opaque in the isotropic phase. Hot spots are then visible as dark stains on the circuit surface.

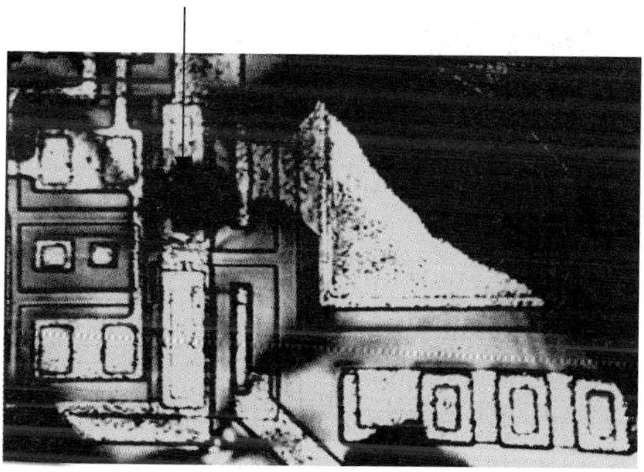

Fig. 63. Photo of an pn-junction submitted to ESD-stress. A droplet of liquid crystal was put on it, the ESD-damage gave rise to a hot spot changing the liquid crystal from neumatic to isotropic phase. The dark hot spot is visible at the top of the arrow in the photo. The spot is very diffuse due to the sensitivity of the state of the crystals for small variations in temperature

Infrared microscopy is based on the thermal radiation (infrared part of the electromagnetic spectrum) from surfaces. Scanning a surface with the microscope and making simultaneous measurement of the radition gives the possibility to detect, if the resolution is sufficient, small, hot points.

A modern method of localizing defects offers *emission microscopy*. In defects there are recombination of electron-hole pairs, created by impact ionization. The photons emitted at the recombination are detected. Photons can also come from "bremsstrahlung" (retardation radiation). The detected emission is amplified and the picture, i.e. one or more luminous points, is superposed on a light-optical picture of the object.

This method is very efficient for detection of most kind of ESD-defects. It replaces simplier, but more tedious ways of analysis, where the surface of the object is studied by microscope in complete darkness. The sites of the defects are also rapidly localized by EMMI. See fig. 64 and 65.

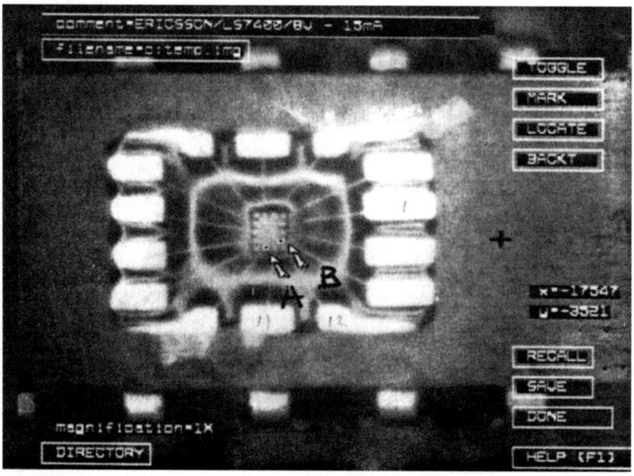

Fig. 64. EMMI-photo of stressed part No 3. Damage at A=pin 10 and B=pin 13. No magnification. Device LS742500 V_{CC}=7V

Fig. 65. EMMI-picture of pn-junction in an input diode of a bipolar LS-circuit stressed by ESD pulse. Magnification 100 times. In s SEM no damaged could be detected

To electrically confirm ESD failures a *curve tracer* tracing the current-voltage relation, is applied. To be able to detect very small changes of leakage current, the "harmonics" of the current are measured, usually the third harmonic. Latent failures (see next chapter) in semiconductors can be revealed by this method (See page 89).

5.5 Cause of failure and failure mechanisms

5.5.1 General

The effect of an ESD-pulse on a microcircuit depends on the pulse height (breakdown power), fast rise-time (fast temperature increase) and energy (developed heat).

The character of the damages varies with the parameters mentioned. Sometimes it is difficult to settle if a component damage originates from an ESD- or EOS-stress. EOS can be a common system transient or of EMP-type (Electro Magnetic Pulse). The latter has a slower rise-time and considerably greater pulse length (µs to ms compared with 1 to 500 ns for ESD-pulses). Usually the protection circuits are first damaged, making the circuits vulnerable for even a minor overstress, i.e. they are eliminated. ESD-damages can however be separated from EOS-damages because they are not extended and commonly occurs near connectors (pins). But an EOS-damage is sometimes developed from preceding ESD-damage.

5.5.2 Common types of defects at semiconductors

The extension and character of an ESD-damage is a function of pulse energy, discharge time and the specific heat conduction of material. See fig. 66 and 67.

Defects use to be characterized as power or current induced destruction or voltage dependent as electric breakdown or discharge. In the former case the result will be melted zones, metal migration. Very strong electrostatic discharges give usually rise to catastrophic failures of the components. The discharge energy can cause melting damages in conductors, ultimately a disruption, or melt the material to short circuits of pn-junctions, or result in a dielectric breakdown in capacitive elements. Such breakdowns were observed at thin gare oxides in MOS-structures.

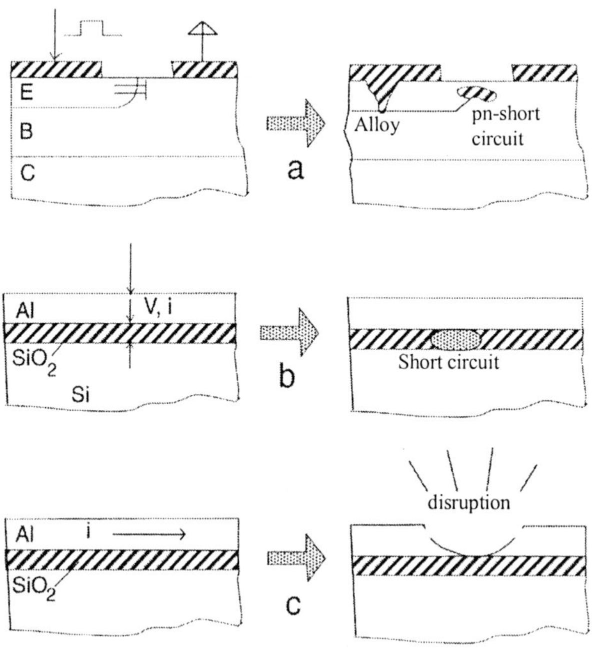

Fig. 66.
a) Overload resulting in short circuit of an pn-junction
b) Oxide breakdown and short circuit
c) Overload of the metalization after current enforcement

While very strong discharges would induce the described catastrophic failures, low level transients are considered to cause "soft" or latent type of failures (see next chapter). In the last case the components demonstrate the "lengthy" type of degradation. Repeated discharges could ultimately transfer the component from a passive state to one with parameters outside specified values or a catastrophic state, where the component is out of function.

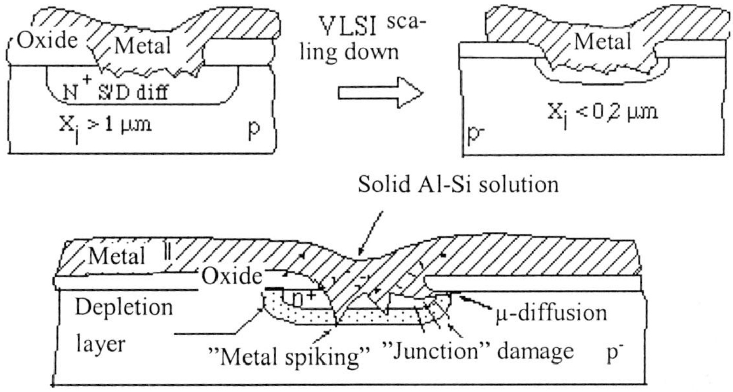

Fig. 67. Aluminium migration. By scaling down monolithic circuits, e.g. VLSI, the diminished diffusion layer can be subject to a breakdown due to so called spikes of metals. Aluminium from metal gates of a MOSFET can also start migrating into the SiO_2-layer. A more acute case is that with an aluminium deposition in contact "windows" immediate above the silicon substrate

5.5.3 Passive thin and thick film components

Passive circuits (resistors) can get the same defects as conductors on IC-components. Especially tantalum-nitride resistors (TaN-resistors) with meander configuration (meander = serpentine shaped) get cracks at "corners" with high field concentration. See fig. 68.

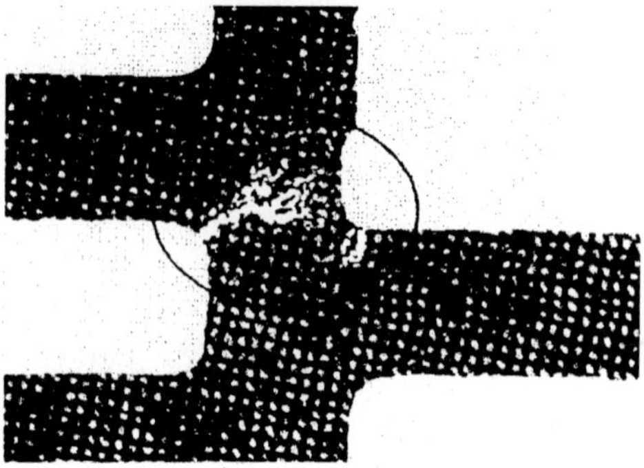

Fig. 68. An TaN-resistor damaged by an 1400 V ESD-pulse of HBM-type. Magnification 450 times similar damages are observed at thick films after ESDstresses.

5.5.4
Example of a simple failure analysis

At the end of the 60-ies ESD-sensitive components were packed in boxes of styrene. This material is an excellent insulator, but it is also easily charged tribo-electrically. When unpacking the component it often happened, followlingly, that the charge passed a pin of the component resulting in an ESD-damage. Protection measures were taken. The following figures are highlighting such a case with ESD-damage and applied failure analysis.

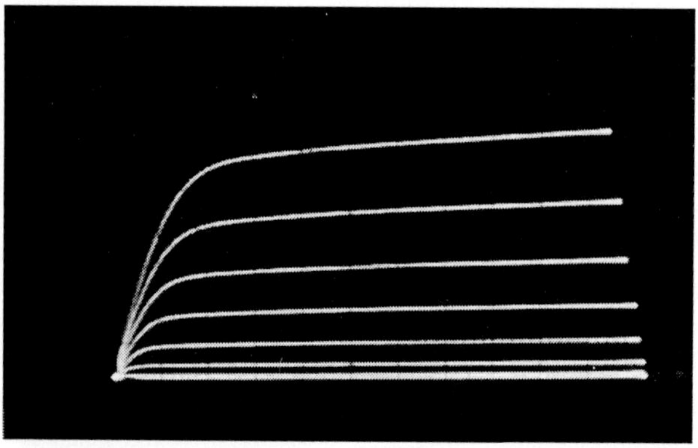

Fig. 69. The characteristics of the transistor is recorded by a curve tracer immediately after unpacking in an ESD-protected environment. It is an N-channel FET, type 2N3797, in depletion mode, i.e. it is conductive at zero gate voltage.

Fig. 70. The characteristics of the transistor after a stress with an 1000 V ESD-pulse of HBM-type. Compared with the characterisistics given in figure 69 the leakage current are increased and during the tracing of the currents a strong current rise appears totally the characteristics.

Fig. 71. Tektronic curve tracer, type 577, used to record the characteristics fig. 72 and 73

Fig. 72. Tool to remove transistor encapsulation, composed by two roller bearings and a cutting trundle.

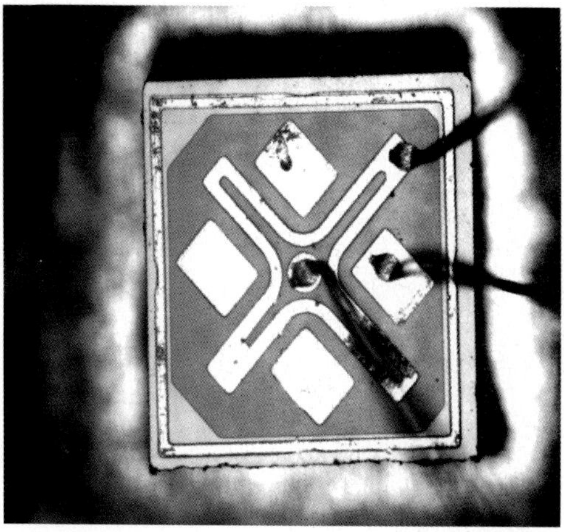

Fig. 73. Investigation on light microscope, so called metal microscope with illumination in the beam passage showing the image of the FET-chip, magnification 100 times. Observe the three bonds to drain, source and gate A corrosion of the gate metalization is visible to the right at the bottom of the picture. This was probably due to unsufficient washing after the manufacturing process. There are other small damages of the metalization, but also two areas, with ESD-damages from the discharge.

Fig. 74. In the center of the silicon chip two damages at the metalization are visible, maybe melt defects and then caused by the ESD-stress. This is difficult to state definitely by the light-optical microscope, though the magnification is 400 times.

Many sophisticated descriptions of failure analysis are found in the literature. See also next chapter.

Finally a more severe ESD-damage is shown, se figure 75.

Fig. 75. FAST circuit stressed with +2000V ESD-pulse, 5 times (1000 x times magnification). One crater formed at clamping diode. Melting process

A special method for failure analysis is "Acoustic Microscopy", that uses ultrasound in the frequency range 5-500 MHz. By this metod a microcircuit can be investigated without removing the encapsulations. Delamentation, voids and wire bond damages are typical defects to be found.

Two different kind of acoustic microscopes are used: The Scanning Laser Acoustic Microscope (SLAM) and the C-Mode Scanning Acoustic Microscope (C-SAM) (Carbone, 1992), (Siemens and Kessler, 1991). An atomic-force microscope (AFM) has also been used (Never and Voldman, 1995).

Suitable samples for analysis of bulk materials can be prepared by the Focused Ion Beam method, FIB (Sudraud, 1991).

6 Latent failures

The concept latent failure related to electronic circuits implies that some kind of damage occurred, but the circuit is still functioning within specification. The word latent comes from the word "latents" = hidden. In this context it means that either the life-time of the circuit is shortened or that it will fail for a minor stress, that would not harm an undamaged circuit. The French expression for latent failure is "defaillance potentielle" (potential failure). This phenomenon has become rather important as results of ESD-stresses, though it can have other causes.

A component with a latent damage is degraded. Often they develop in intermittent failures and eventually to catastrophic failure. The appearance of this procedure, to degrade by time without additional stresses, was discussed very much. Failure because of additional minor stresses has been confirmed clearly on experimental level. The first systematic investigation to confirm the existence of latent failures was presented in 1982 (McAteer and Twist, 1982).

It is difficult to get an idea about the field failure rate depending on earlier latent failures. Failed components must be collected from the field and submitted to failure analysis to detect those with ESD-failures and ultimately the failures preceded by latent failures. Failure analysis is time consuming and costly, consequently only few extensive investigations were performed.

The number of latent failures given from such investigations vary from 1 to 50 %.

A literature search on investigations of latent failures was distributed at the 1987 EOS/ESD-symposium. In 42 of them latent failure were confirmed and only 8 investigations had given negative result. The list was not complete and latent failures are still the object of much component research. Attention must be paid to this phenomenon in electronics!

6.1 Time dependence and event dependence

Those latent failures degrading by time to catastrophic failures are called "time-dependent" and those transformed into catastrophic failures by minor additional stresses are called "event-dependent". The latter are recognized while the others are questioned. Filed failures are often preceded by latent failure in manufacturing. The development of an event-dependent latent failure to catastrophic failure, for instance damage by melting, is likely to proceed rapidly without any sign of malfunction as intermittent failures, functional faults etc.

There are indications about existence of the time-dependent failures. Laboratory tests with artificially stressed components show that catastrophic failure can occur by temperature-acceleration of them. These investigations have, however, been critcized owing to component choice, testing conditions and especially, because other investigations gave no evidence for time-dependent latent failures. Maybe the explanation is, that this kind of latency develops very slowly to full failure, i.e. a somewhat early ageing process.

6.2 Technology dependence

6.2.1 CMOS-Technology

Latent failures in CMOS-components are hard to indicate. This is very much due to the strong scaling down of this technology, with submicron metalizations, giving small marginals for overstress, charges injected in the oxide can cause electrical breakdown (Wolters and Zegers-van Duyshoven, 1987). The process is analogue to measurement of materials by small stresses to rupture. See fig. 76.

Investigation of MOS capacitor

Two selected MOS-capacitors were submitted to ESD-HBM-pulses. The first one was stressed by successively incremented voltages, 100 to 500 volt, and measured between the discharges. For the second capacitor repeated 200 volt pulses were used, again with measurement between them.

Fig. 76. EBIC-photo of stressed MOS-capacitor with triple damage from one pulse

Using a scanning electron microscope with EBIC-mode, damages in the capacitor were revealed (EBIC = Electron Beam Induced Current) CV-curves indicated a defect increase in the SiO_2/Si-interface and by a third harmonic measuring instrument the formation of latent failures was detected, see next table.

Stress level	Third harmonic (amplitude of basic frequency, 20 mV)	
	Component 1	Component 2
100 V	2.6 μV	
200 V	1.6 μV	1.4, 2.2*, 30** μV
300 V	1.7 μV	
400 V	1.5 μV	
500 V	3.2 μV*	

*) Latent failure
**) Latent to catastrophic failure

A latent failure of the first component is introduced after fifth pulse (500 V). In the second component the initial pulse resulted on a latent failure after two pulses, that developed into a catastrophic damage at the third pulse. Failure was less in this case. In this way it was possible to follow the development from latent to completed failure. See fig. 77 for detection by CV-meausrement.

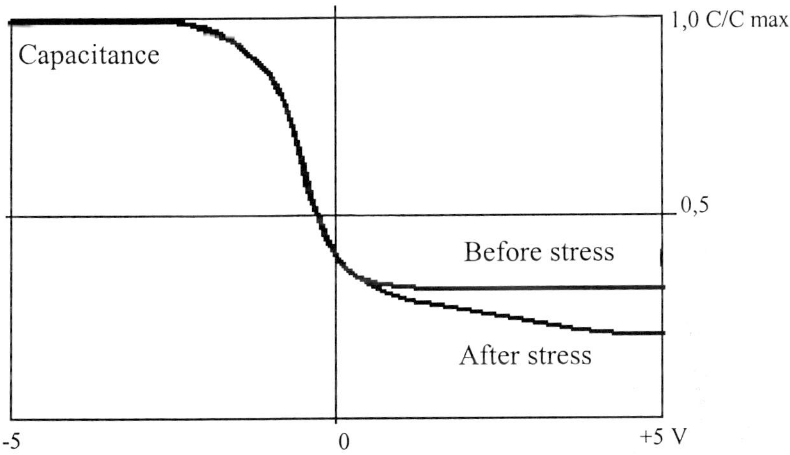

Fig. 77. CV-curves of MOS-capacitor. The voltage across the capacitor is incremented from-5 to 5 volt, the capacitance being measured simultaneously. The presence of latent failures are revealed by the measured change of capacitance

Bipolar silicon microcircuits

Especially the old types of bipolar components were proved to get latent failures by metal-diffusion into pn-junctions, to the insulation layer or the interface between them. The process can start by a break through opening a way for higher current density. The area discharged by ESD is not stable as approved by the "healing effect", i.e. an apparent recovering is obtained if the component is left inactive for some time.

6.2.2
Investigation of Low Power Schottky TTL-components

Because the existence of latent failures had been questioned very much, an investigation was performed to verify this phenomenon of latent failure (Hellström and Welander, 1991). See fig. 78.

For the experiment 74LS00 TTL-components were chosen. They were initially stressed by HBM pulses, using the leakage current as damage indicator. After the stress they remained inactive and protected against ESD for a whole year, figure 78. No substantial change in leakage current was recorded during this period. After half a year more same components showed an increase leakage current. The following dynamic test gave interesting results. A testing temperature of 125°C was applied for 21 pieces, of them 7 had showed sign of latency before the test, i.e. leakage change less than to specified 20 μA value (20 μA = acceptance value).

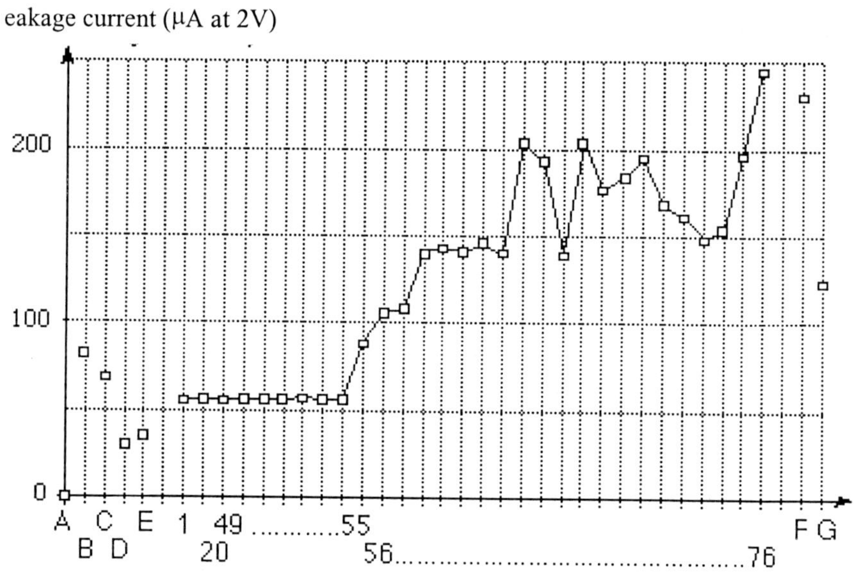

Fig. 78. Time story for an ESD-damage component (74LS00)

A: 2 April, 1985, undamaged component
B: 3 April, 1985, exposed by person charged to 1000 V
C: 11 July, 1985, healing-effect after baking at 125°C
D: 28 August, 1985, continued healing effect at check measure
E: 1 April 1986, increasing leakage current after further storage
Day 1 (17 Dec, 1987): Control measurement and preparation for dynamic test
Day 2...55: control measurements, small changes at the end of the period
Day 56: large increase of leakage current
Day 57...75: changes, mostly increasing leakage current
Day 76: (3rd March, 1988): end of dynamic test
F: 10 August, 1988, control measurement after relaxation in storage
G: 1st Dec, 1988, control measurement, the investigation closed

After about eight weeks the leakage current began to pass the 20 μA level for three of the seven components. For eleven of the other components, the leakage current started to increase at the same time. Their currents were > 20 μA from the beginning. This is an important observation as they had functioned properly with the increased leakage before the dynamic test! Compare with fig. 79.

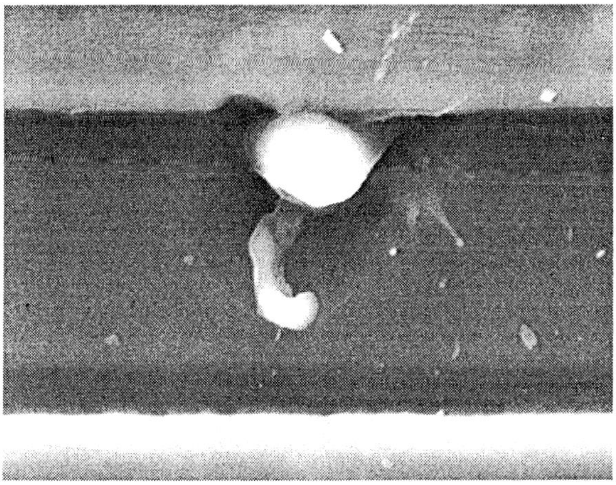

Fig. 79. SEM picture with 7500 magnification.
To investigate what value the leakage current should have to make detection of an ESD-damage possible the following procedure was used. Some non-stressed components of same kind as before were exposed to ESD-pulses of different height. The leakage currents were measured afterwards and a picture of the damage taken. The picture above with the dendrite-like formation of silicon and aluminium was made after an ESD-pulse that resulted in a leakage current of 23 mA, i.e. 1150 times the accepted value. This illustrated the difficulty to detect latent failures.

The result of the investigation is very important when comparing with the negative result of another investigation (Woodhouse and Lowe, 1988). It is possible that the accelerated tests in the latter case were finished to early. Tests at

25°C performed at the same time as at 125°C in the described investigation seem to confirm that. In this case nine had latent failures before the dynamic test, but their leakage current did not increase during the first time. On the contrary components with currents > 10 µA before the test, got a further increase, as in the 125°C test! The time for the dynamic test was probably too short to influence components with latent failures at 25°C. A prolonged dynamic test is likely to have increased the leakage currents even in this case.

6.2.3 Bipolar ECL

In computer logics ECL-circuits are very usual. Unfortunately they get rather easily latent failures and can cause troubles. Investigations in Japan showed that "unexplainable" failures of computer systems after some time in use probably had ECL-components as origin.

The damage started by a diffusion of metal into the base-emitter transition. The electrical change in the circuit is represented by a parasitic resistance, R_p between base and emitter, figure 80.

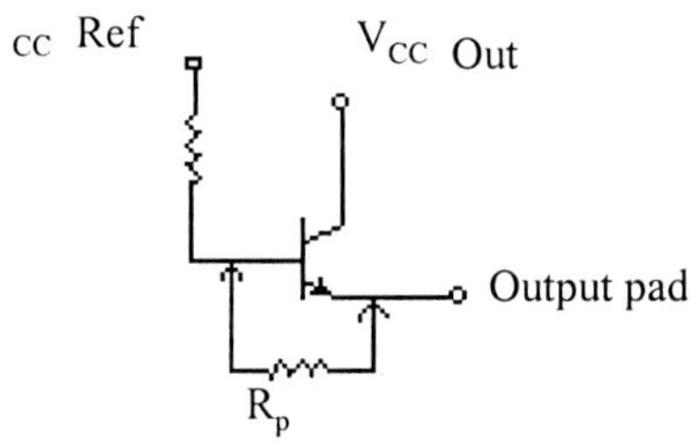

Fig. 80. Equivalent scheme of the destroyed ECL-output

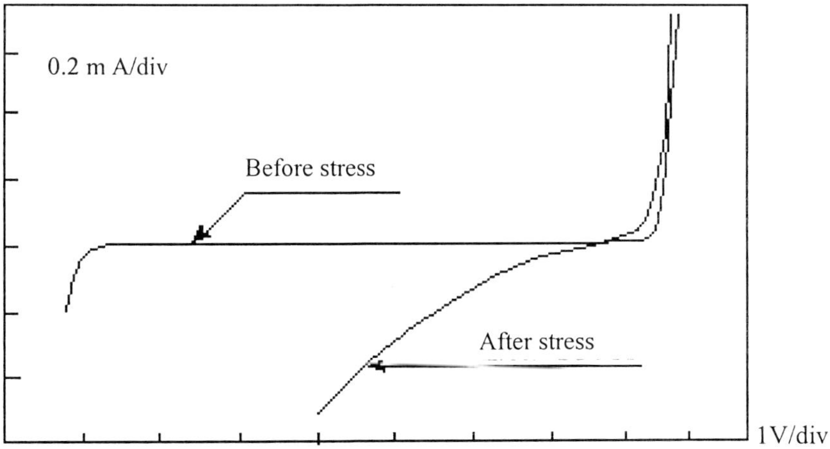

Fig. 81. I-V-characteristic of an ECL-circuit, before and after an ESD-transient

Damage can be detected by the I-V-characteristic change, demonstrated in figure 81. By time the latent failure develops to a catastrophic breakdown. The damage is observable in a SEM-microscope, figure 82.

Fig. 82. Picture of ESD damage of the output base-emitter transition in an ECL-component

In an investigation of ECL-microcircuits third harmonic measurements were performed. It was possible to follow the development of the damage from latent to complete failure increasing the third harmonic. This method is ten times more sensitive than direct recording changes in the I-V-characteristic, what was done for comparison. Accelerated tests by a chock-chamber (rapid temperature change) confirmed the presence of latent failures. Several components with small changes after ESD-stress failed in the stress, while unstressed samples remained unchanged.

6.3 III-V-COMPONENTS

A case with semiconductor lasers to be described later showed that III-V-components are prone to get latent failures. However, certain MESFET-GaAs microcircuits proved, that they do not degrade successively after any minor stress but fail catastrophically at some ESD-level. This speaks for a strong dependence on technology as to formation of latent failures.

Another example concerned qualification of a LED component (LED = Light Emitting Diode). The first measurement revealed increased leakage current but within specified values. The component changed in the life-time accelerated test, the leakage current increased above accepted value. Also other parameters were outside specified values! This case is a rather strong indication for III-V-components to get time-dependent latent failures. Optocomponents are often constructed with these materials.

6.3.1 GaAlAs-lasers

Already in 1984 an investigation was presented about latency in GaAlAs-lasers (Sim et al, 1984).

Semiconductor lasers are known to be sensitive to ESD. The referred investigation was based on the experience that handling caused ESD-failures, thus the HBM-method was applied. In the experiment a current threshold change of 3 mA was used as indication of catastrophic failure.

The goal implied to determine the transient-amplitude introducing latent failures with successive degradation. Test equipment consisted of capacitors in the range 120 pF to 16 μF, that were charged and discharged through the lasers using a mercury switch for charging and discharging.

With "amplitude" was meant the peak value of the current pulse, the width of which measured as decay time to half the peak value, 50 ns-1 ms. Pulse shape could be carefully registered by oscilloscope using probes for currents of 12 A.

The characteristics, pulsed light/current, was traced before and after ESD-transient. The amplitude of the transients was increased at 25°C until the characteristics changed. Fifty lasers were studied for detection of latent failures. Twenty-five of them had already passed a life-time test for 12000 h at 70°C, with established degradation level.

With the aim to reveal latent failures a burn-in process at 70°C, with 5 mW optical effect for at least 1000 h preceded the ESD-test. This gave the normally expected increase of the threshold current at 25°C. Light current curves were traced repeatedly before and after ESD-transients. The light output of 5 mW was regulated by the current to remain constant during the tests. The pulse width varied from 50 ns to 160 μs (decay-time) and the amplitude did not pass 1/10 of the value registered to have resulted in an immediate change of the laser output. At measurements performed, different effects of the transient appeared and latent failure could be detected in the samples submitted to thermal acceleration after test with minor ESD-transients!

A failure analysis of stressed parts revealed the failure mechanisms. Transients in forward direction created heat, lowering the band gap, what produced more heat. Eventually melting occurred. Failures from pulses in the reverse direction should have the origin in locally melting of the laser material caused by avalanche breakdown. The result of this investigation supports also the technology dependence.

6.4 Polysilicon resistors

The creation of latent failures in polysilicon resistors is more pronounced than in diffused resistors. This is probably due to the less efficient heat conduction from polysilicon resistors. See fig. 83.

Many input protection circuits have polysilicon resistors. The latent failures can erroneously be ascribed the technology of the component. Careful analysis needs to localize the site, where the first signs of failure appear and the pertinent technology! An input protection circuit of an CMOS-component consists commonly of a polysilicon resistor to limit the current and two diodes that reduce high transient voltages (chapter 11, Protection methods).

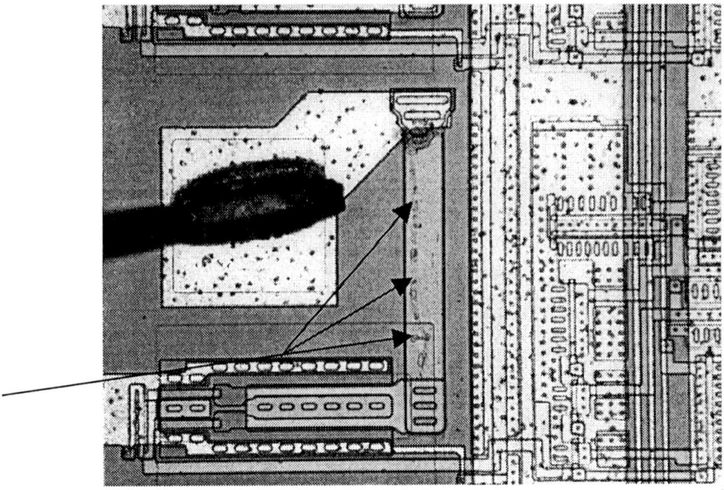

Fig. 83. Polysilicon resistors as part of a protection circuit showing beginning melting damages, the stripes pointed by the arrows

6.5 Detection of latent failures

The failure mechanism of a latent failure can be the beginning of a "bridge" between two conductors (increased leakage current), or first structure change for oxide breakdown, partly melted conductor or metal diffusion into a pn-junction.

Current voltage dependence (I-V-characteristic)

A degradation or damage of a component can normally be detected by measurement of current leakage in the reverse direction using for instance a curve tracer. The leakage currents caused by latent failures are often so small that they area not observed or merely neglected. A way to record the presence of latent failures is to use the fact that for some part of the I-V-curve for an undamaged component the slope shall be zero. Latent failures will change the slope somewhat.

Measuring the slope between two points is a more sensitive method to record changes than a direct measurement of the leakage current. Another advantage by measuring the differential $\Delta i/\Delta v$ is the elimination of possible bias currents (if they remain constant), fig. 84.

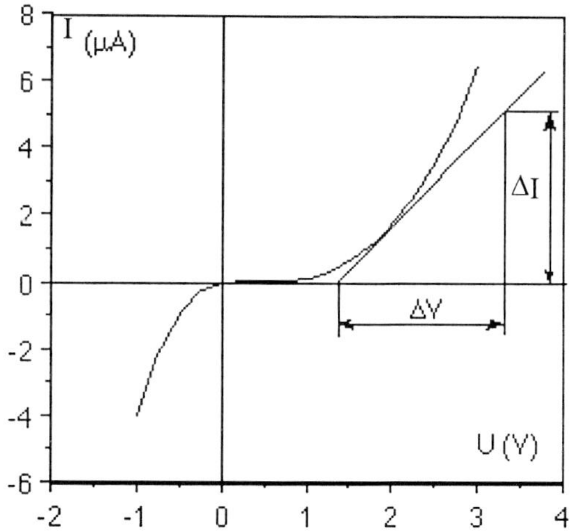

Fig. 84. Measurement of changes in the slope of the I-V-curve is a more sensitive method than a direct measurement of the leakage current and elimination of bias currents is important when measuring small currents.

6.5.1 Third harmonic method

Another possibility to detect changes of the I-V-characteristic offers measurements of non-linearities, fig. 85.

A pure sine shaped current through a resistor, capacitor or inductive element containing non-linearity will create harmonics in the voltage drop. The third harmonic has proven to be most significant. A commercial instrument has been developed, having a bias frequency of 10 kHz and selecting 30 kHz, CLTE (Component Linearity Test Equipment) from RE Instruments AS, Copenhagen.

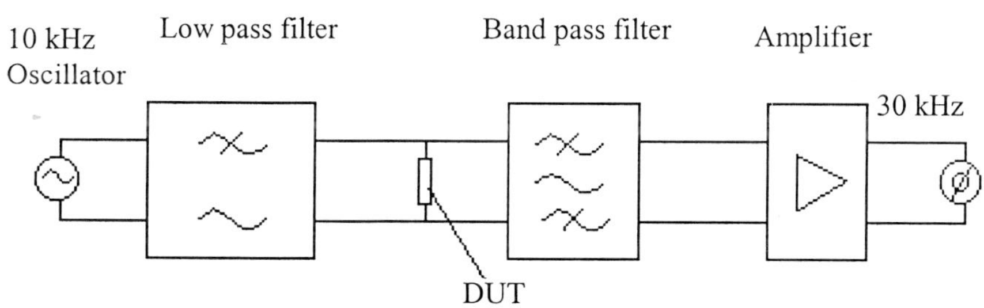

Fig. 85. Simplified scheme of CLTE, "harmonic" instrument

6.5.2
Quiescent current method

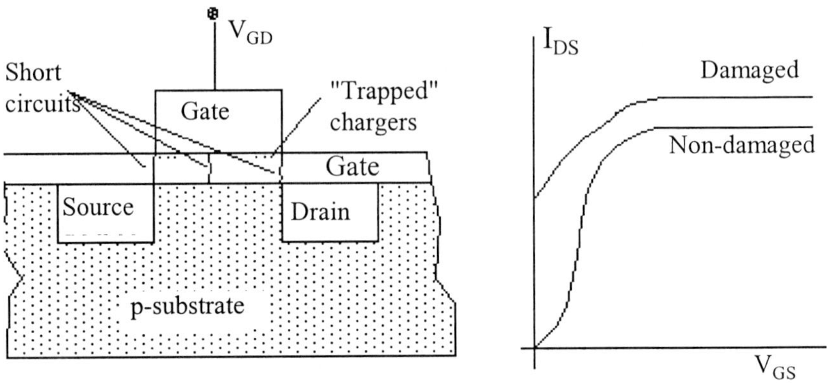

Fig. 86. Typical failures in MOS-structures

In the quiescent current method the feed of current is measured, the component being in a quiescent mode. This method is very effective for detecting defects in circuits of CMOS technology. The current is normally less than 10 nA.

As seen from figure 86 the quiescent current is increased in circuits with defects type "charge traps", gate oxide short and parasitic transistor leakage. But the logic states of the outputs remain intact.

6.5.3
Capacitance-voltage measurement

The CV-method offers a powerful test. The measurement can reveal doping profiles of materials settled in the oxide. The characteristic shape of the C-V-curve depends on electrical stress, damages, annealing degree or interface traps. Figure 87 shows how the information is obtainable from a capacitance measurement at a high frequency and quasistatically as function of applied voltage characteristics.

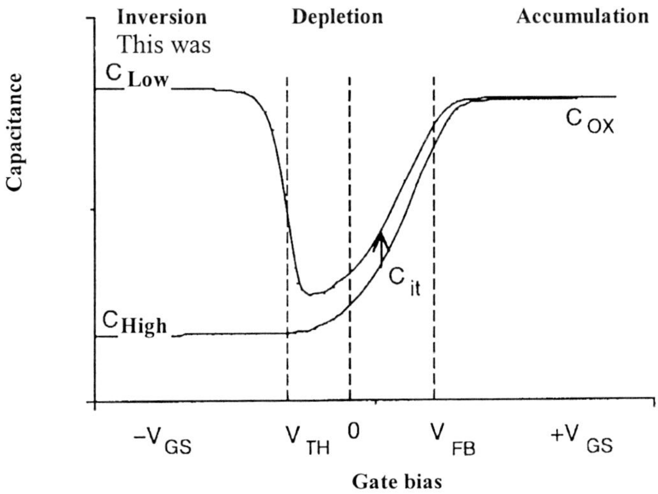

Fig. 87. Typical CV-characteristic of n-type substrate capacitor

The quasi-static capacitance value must be measured at a frequency, low enough (mHz) to get a complete response from all "traps" to the "step-shaped" voltage. The measurement of the capacitance at the high frequency is achieved by superposing the step-shaped voltage with a low-amplitude (10mV), sine -shaped signal of high frequency (100 kHz or 1Mhz). It can be assumed that all "traps" make a complete response to one voltage step.

6.5.4 Image processing

Detection of latent failures is also possible by direct image processing. Damages that can be discovered are small permutations of the material structure from first initial breakdown in the component. Images of exposed components are iterated backwards to find the level at which certain defects as voltage breakdown, material transport (migration) or melting occur.

Image processing implies storage of a picture in a computer as a signal and to perform digital operations with this. Different "operative modes" are used on this purpose, like amplifiers, indicators, comparators and filters. "Hidden" properties, e.g. emphasing of lines, can be found. The method is time consuming and it claims good experience to be able to apply the proper operator to enhance the wanted properties.

Image processing of a FAST-circuit

The described method was used on pictures of ESD-exposed FAST-circuits. Iterations backwards gave the susceptibility level. After a 500 V-pulse some weak signs of damage could be seen in a SEM, figure 88.

The pictures were stored in an image-processing computer. First treatment was to eliminate noise by a low-pass filter. To demonstrate patterns in the picture a vertical enhancement operator was chosen. The operator "amplifies" vertical lines and suppress horizontal lines. A similar horizontal operation gave negative results, no particular patterns came out. See fig. 89.

Fig. 88. FAST-circuit stressed 5 times with 500V-pulses. A weak sign of damage on the clamping diode is observable. A SEM-picture with 1000 magnification. The damage is with a "ring"

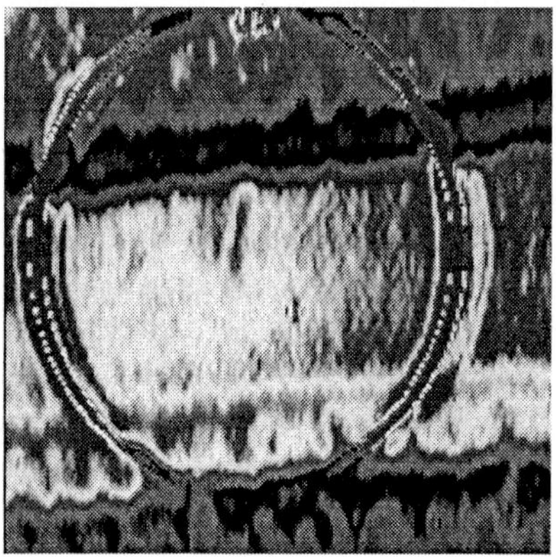

Fig. 89. The same picture as in figure 88 but after a twofold processing in vertical mode.

After the image processing a "channel" covering one third of the distance anode - cathode is visible. There are additional thin lines forming a "bunch"-like pattern from the channel towards the cathode, fig. 89. Traces of aluminium in the channel were found by material analysis. The picture obtained gives a better understanding how the failures develop from a first breakdown to completed catastrophic failure.

6.6 Consequences and counteractions

6.6.1 Marginal until malfunction

Latent failures are very difficult, not to say impossible to discover at components in field use. The definition itself of latent failures declares that failure can have been initiated but will not be evident until a complete failure comes. Not only ESD need to be the cause of latent failures. In the manufacturing process small changes of the component can be introduced by dislocation or material contamination. The processes of the component material give a normal ageing and in the long term develop damages similar to the result of a few ESD-transients on a quite new component.

Normal applications of components are usually not associated with stress. Components are designed and manufactured to get good "marginals" to failure at normal loads, i.e. stresses are rather low. Under such conditions the component life-time is counted in millions of hours.

A latent failure will diminish the components marginal to failure, i.e. cause too early destruction of the component.

6.6.2 Investigation of marginal to malfunction

A manufacturer investigated the marginal to malfunction for one of his components by studying its behaviour at different current leakage. The components were divided in two groups according to their leakage current. One group having 0.1-0.2 μA respecitvely 0.6-0.8 μA as new components, approved leakage being 1 μA. For these components a burn-in at 150°C for 24 hours was performed.

For the first group the leakage current was unchanged after burn-in, while about 5% of the second group got leakage currents above 1μA! After that experiment the approval level of component leakage current was set at 0.5μA! Components with leakage near 0.5μA must still be regarded to have a diminished marginal to failure and are likely to have latent failures!

6.6.3 Consequences

Components with latent failure have turned less reliable. Often failures have an intermittent character before a catastrophic failure, giving a prewarning. The event-dependent latent failures are insidious, causing catastrophic failures without any prewarning. Minor transients from the transmission line can destroy components with latent failures.

Latent ESD-failures occur most often after delivery from the vendor, because the environment is usually very well protected against ESD in production and testing. The customer might sometimes have incomplete ESD-protection.

6.6.4 Counteraction

As latent failures are insidious and not detectable in a simple way the components must be protected against ESD events as efficiently as possible. A careful attention to parameter changes can help reveal perturbations. Components are to be handled according to a well elaborated ESD-program, pertaining to the normal quality actions of enterprises, acting by a realistic TQM-philosophy. See also chapter 11.

7 Indirect ESD

In the preceding chapters phenomena associated with electrostatic discharges into or inside circuit elements (direct ESD) were treated. Damages are caused by currents (diffusion, melting) or voltage (breakdown).

According to the eletromagnetic theory an electric breakdown through an airgap gives rise to an electromagnetic pulse, wave-package, i.e. a limited number of waves are emitted, see figure 90. This occurs for instance, when a charged body is brought close to a metal or earthed conductor. The same happen when a charged person is seizing a door-handle or some conducting part of an equipment, fig. 90, 91.

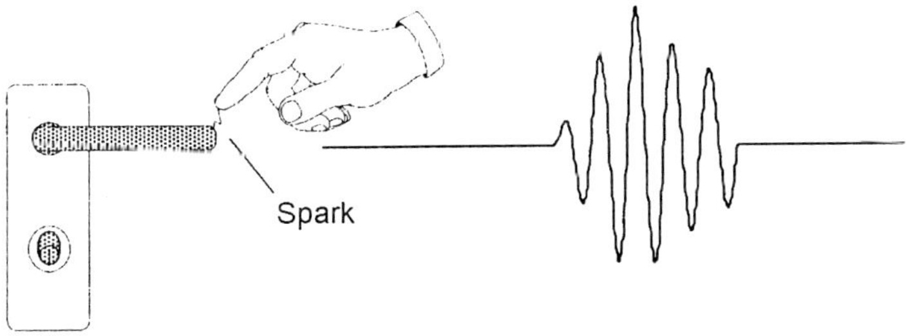

Fig. 90. Discharge occurring on touching conductors, with "pulse-train", ("wave package")

Fig. 91. A person being charged can unintentionally cause a discharge to a cabinet with a spark. If there are memories they may loose information - even get damaged

By this we are touching a new phenomenon of great importance from the view of electrostatic perturbation. The first observation of it was done as long ago as the end of the last century. But triboelectricity was observed already in ancient time.

The German physicist Heinrich Hertz performed the first successful experiments with wave propagation from a spark discharge, fig. 92.

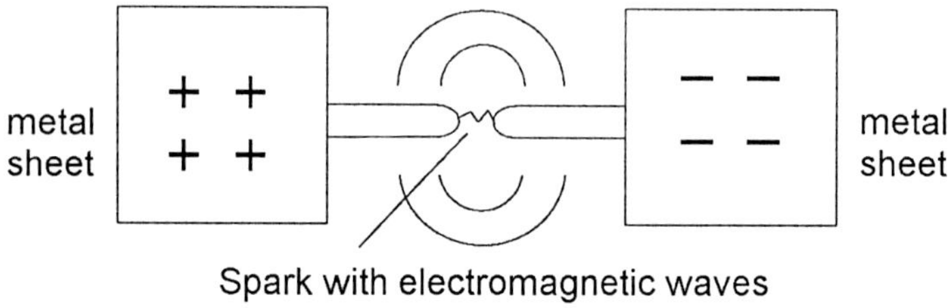

Fig. 92. Approximate arrangement of Hertz's trials

These waves were also called "Hertzian waves". The existence of such waves was predicted by J.E. Maxwell's electromagnetic theory. Led by this Hertz performed his experiment to confirm the theory. Thus the possibility of wire less communication was given. The Italian Marconi made fundamental experiments on this and succeeded to establish wire less communication. The first wire less telegraph utilised spark discharges and was also called "spark transmitter". Hertz's rods, figure 92, were the embryos to modern antennas (aerials) radiating or receiving continuos electromagnetic waves from oscillating charges. The unit for electromagnetic frequency was named after Hertz. 1 Hz = one period per second. Perturbations in the function of electronic systems from the described phenomenon are called "indirect ESD", and difficult to avoid. Frequencies involved amount from 1 MHz to 6 Hz. A complication is that electromagnetic waves have two components, one magnetic (H) and one electrical (E). Mathematically they are represented by two vectors perpendicular to each others and the direction of propagation (transverse waves). The two components have different penetration abilities through matter.

7.1 E- and H-field

According to the Maxwellian theory the electromagnetic waves consist, as already mentioned of a magnetic and electric component passing through vacuum with the speed of light. They are of the same nature as light but much lower frequency, i.e. longer wave-length.

To describe the behavior of the electromagnetic field the distant to the source is important. We define as "near-field" the field within about one wave-length distance from the source. The following relation is valid in a far-field:

$$\frac{|E|}{|H|} = Z = \sqrt{\frac{\mu_0}{\varepsilon_0}} = 120\pi = 377\Omega \qquad \text{(in vacuum).}$$

|E| and |H| means "absolute" values (magnitudes), always greater than zero. Z is called the wave impedance of space, fig. 93.

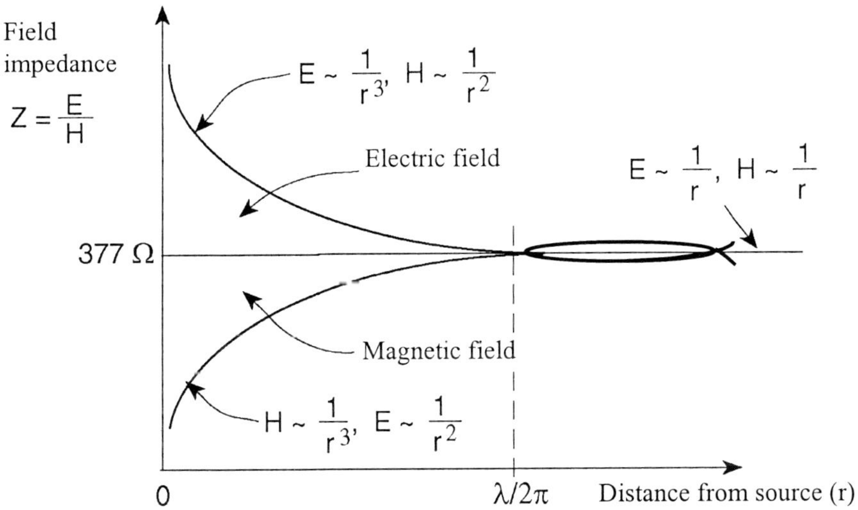

Fig. 93. The electric E-component and the magnetic H-component variations with distance from a source, size much less one wavelength

In the "near-field" the relation between E- and H-components is complex. They can be measured in a "TEM"-cell (TEM = Transverse Electro Magnetic). Antenna probes most sensitive to B- respectively H-fields, are used, fig. 94.

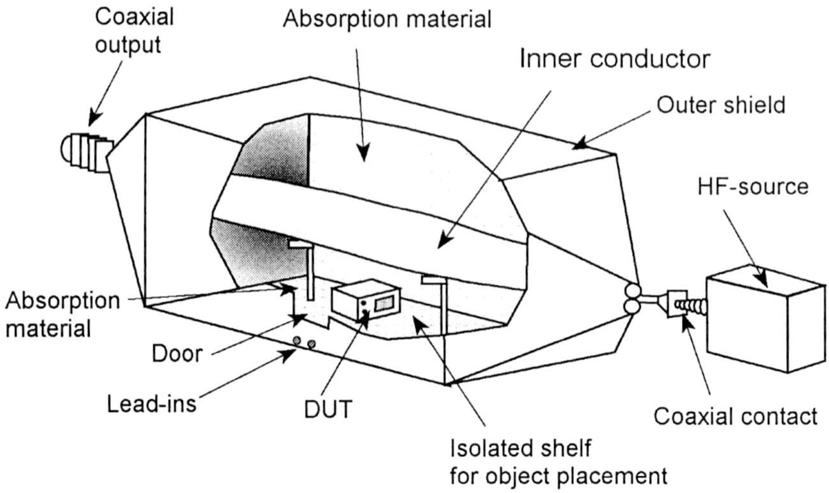

Fig. 94. Schematic picture of "TEM"-cell

The far-field is usually measured by means of a biconical antenna (fig. 95) in the frequency range, 0-300 MHz and a log-periodic antenna (fig. 97) in the 200 MHz to 1 GHz range.

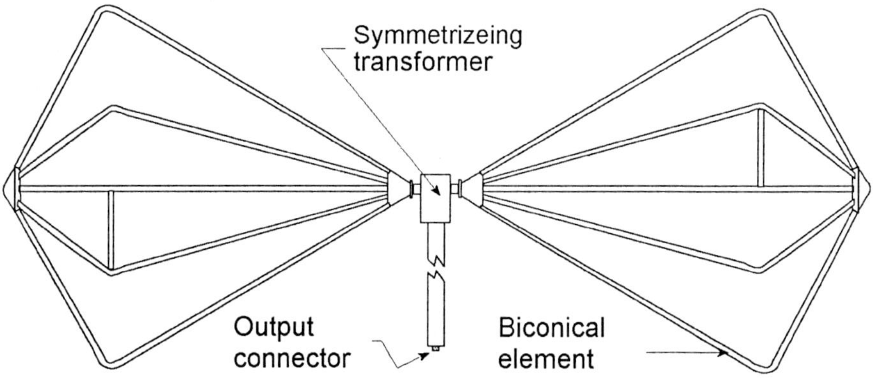

Fig. 95. Biconical antenna

7.2 Examples of perturbation

There are numerous examples how electrostatic discharges in the vicinity can disturb electronic equipments. This is a problem in computer rooms and telephone stations for instance if not a sufficient shielding or other means of "attenuation" are applied. An illustration example will be given. A personal computer (PC) was disturbed functionally by a charged metal chair moved towards another metal chair with a discharge as result, fig. 96, 96 and 98.

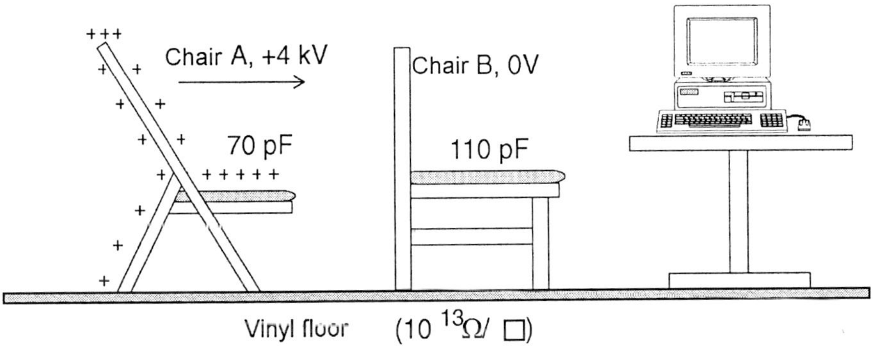

Fig. 96. Generations of EMI by indirect ESD. In the original arrangement chair A was moved towards chair B

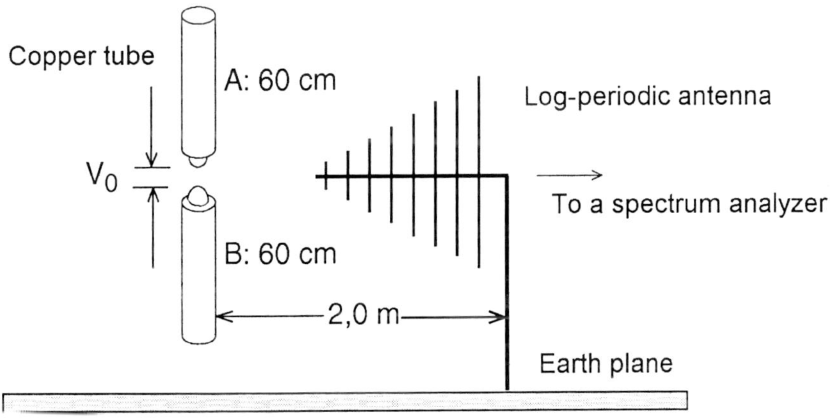

Fig. 97. Measurement of field spectrum for ESD-simulation of the example above

By using a copper box with an opening and making a spark discharge between two copper rods the case with the computer was simulated fig. 98.

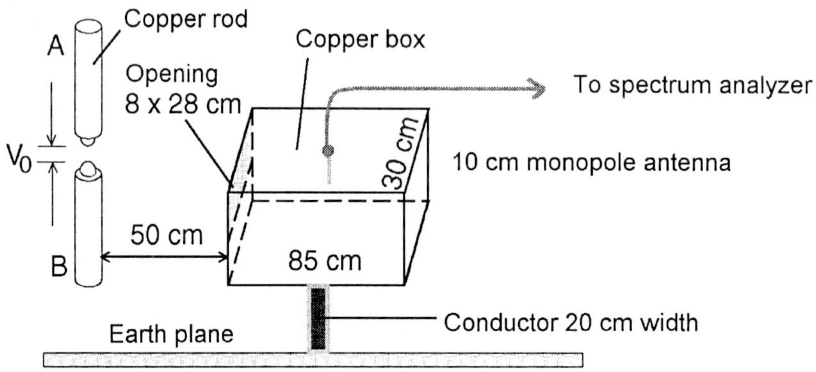

Fig. 98. Simulation of the original arrangement with recording of field strength and distribution "inside" the box ("computer")

Wave spectra were registered both from the spark discharge and the electromagnetic field inside the box, figure 98. Measuring of field spectra was performed according to fig. 98 (Honda and Ogura, 1985). See fig. 100.

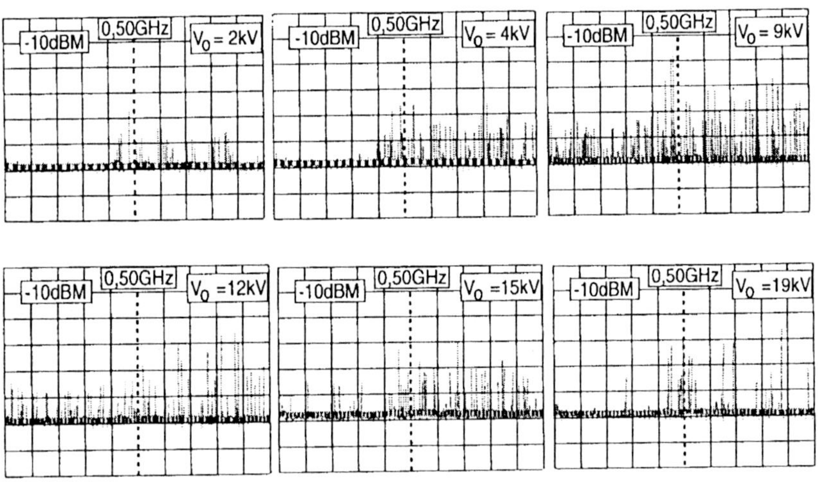

Fig. 99. Recorded spectra inside the box after discharges of 2, 4, 9, 12, 15 and 19 kV (100 MHZ/div and 10 dB/Div)

As seen from the spectra no increased level is obtained above a certain ESD-amplitude. This is due to the fact that a greater part of the energy is spent in the air gap at higher voltage. At discharges with spark gaps to a cabinet as shown in figure 104 the result can be more severe with a 8 kV than 15 kV pulse of the same reason.

A classical problem is the charge partition between two capacitors, next figure.

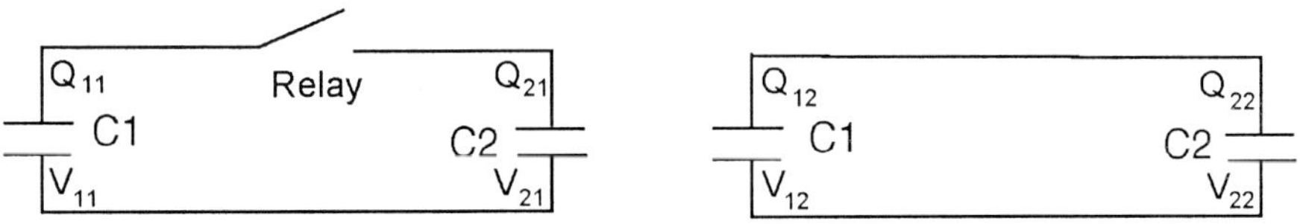

Fig. 100.

a) Starting point
C_1: $Q_{11} > 0$ $V_{11} > 0$
C_2: $Q_{21} = 0$ $V_{21} = 0$

b) After closing the relay
C_1: $Q_{12} > 0$ $V_{12} = V_{22}$
C_2: $Q_{22} > 0$ $V_{22} = V_{12}$

It is easy to believe that the total energy stored by the capacitors will remain unchanged, i.e.

$$E_1 = \frac{V_{11}Q_{11}}{2} = E_2 = \frac{V_{12}Q_{12}}{2} + \frac{V_{22}Q_{22}}{2}$$

But that is not the case and $E_2 < E_1$ because of the energy spent in the relay gap, that again depends on the voltage (Honda 1984, 1985). The process is further clarified by the pictures 100 and 102. A charged capacitor is discharged to earth through an air gap. The energy is partly spent in a resistor, partly in the air gap. The partition depends on the voltage drop in the air gap. This can be demonstrated by measurement of the range of the emitted EMP-wave (EMP = Electro Magnetic Pulse).

For the detection of electromagnetic pulses small detectors are available. They have exchangeable antenna probes of varying length. In fig. 102 a Japanese detector is shown.

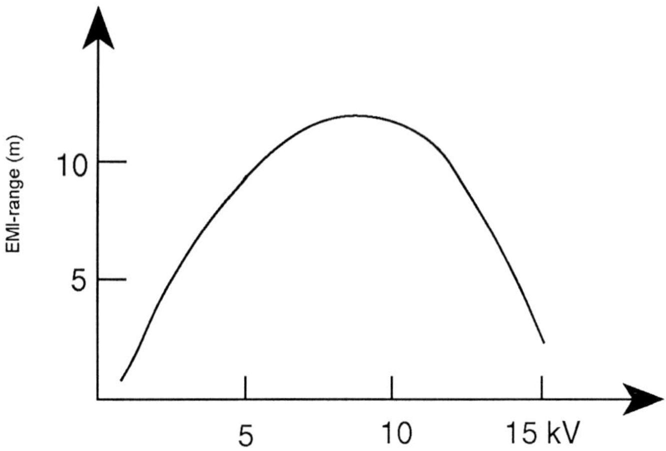

Fig. 101. EMI-range as function of discharge voltage. The discharge across the air gap is equivalent to the discharge from a capacitor of 150 pF with a serial resistor of 150 Ω. With an ESD detector like in figure 98 the max range of the emitted electromagnetic pulse is measured

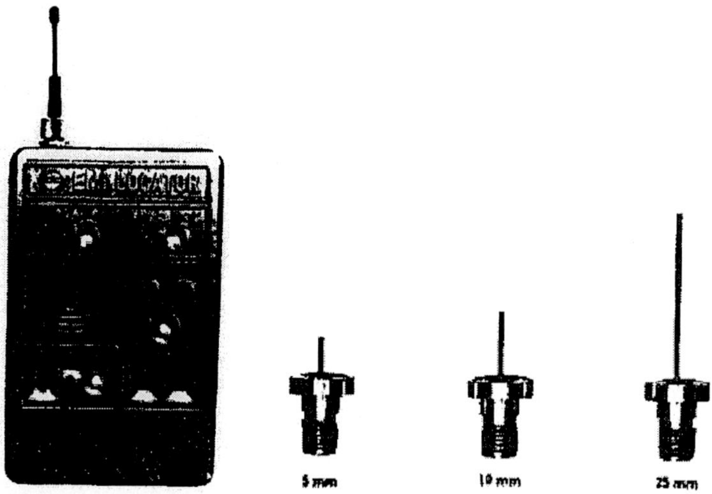

Fig. 102. A commercial Japanese ESD-detector with three short monopole antennas

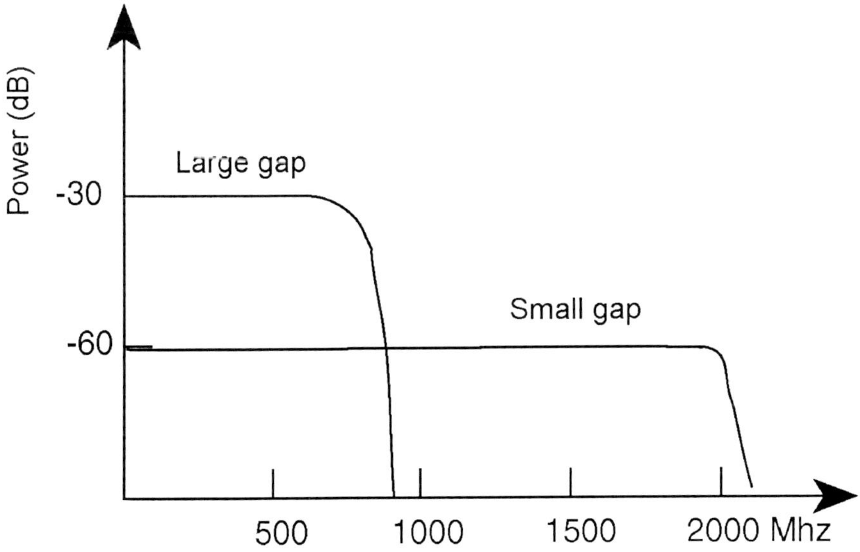

Fig. 103. Transmitted frequency spectrum as function of spark gap. The figure shows that with larger spark gap (and higher voltage) the pulse will have a higher power but a narrow frequency band width than for a discharge in a smaller spark gap and with lower voltage

7.3 Testing methods and norms

7.3.1 EN- and IEC-norms

For EMC (Electro Magnetic Compability) and Indirect ESD the following norms are available (pr = preliminary). EN = European Norms.

7.3.2 Standards (Important for Information Technology)

pr EN 55024-1	General requirements on immunity
pr EN 55024-2	Endurance against electrostatic discharges (IEC 1000-4-2)
pr EN 55024-3	Endurance against radiation perturbations and electromagnetic fields
pr EN 55024-5	Endurance against transients
pr EN 55024-6	Endurance against perturbations from the power line

EN 55022 Radio perturbations - bounding values and measuring methods

Testing according to IEC for information technology equipment (IT) 1000-4-2 (EN 55024-2)

In pr EN 55024-2 there are suggestions for test performance at system level as well for indirect as direct ESD-endurance. On this purpose an ESD-generator outlined in figure 104 is used for both direct and indirect ESD-test.

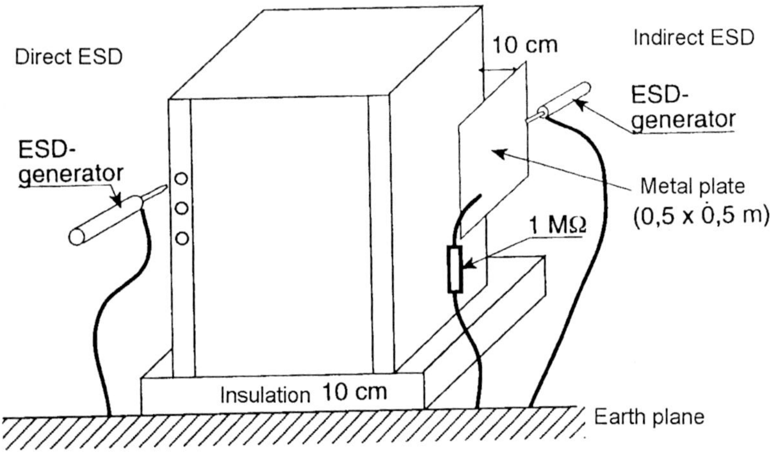

Fig. 104. Testing of an apparatus for both direct and indirect ESD. At discharges to the metal plate this will act as an antenna for the electromagnetic pulse that radiates towards the equipment. Because of the "skin" effect only the surface of the plate is active

The IEC 1000-4-2 norm stipulates at injection test voltages of 2, 4, 6 and 8 kV, possibly higher and if an air gap is present 2, 4, 8 and 15 kV, possibly higher. Injection is the normal method as reproducibility will be difficult if an air gap is introduced. See scheme fig. 105.

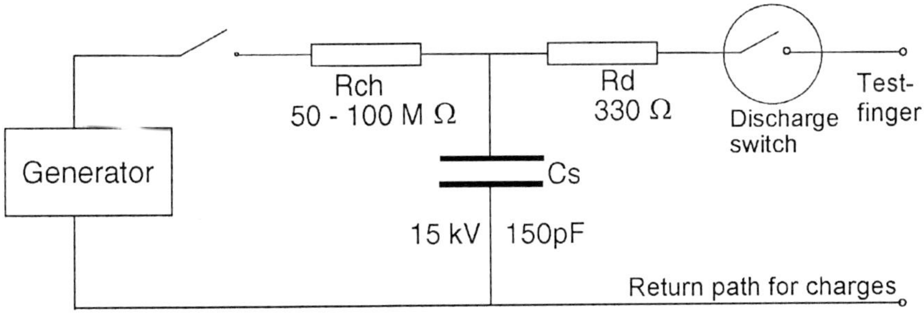

Fig. 105. Electric circuit scheme for the ESD-generators of pr 55024-2

The ESD generator according to pr EN 55024-2 is often designed as a pistol, fig. 106. Calibration procedure is given in the norm. Common pulse form in fig. 107.

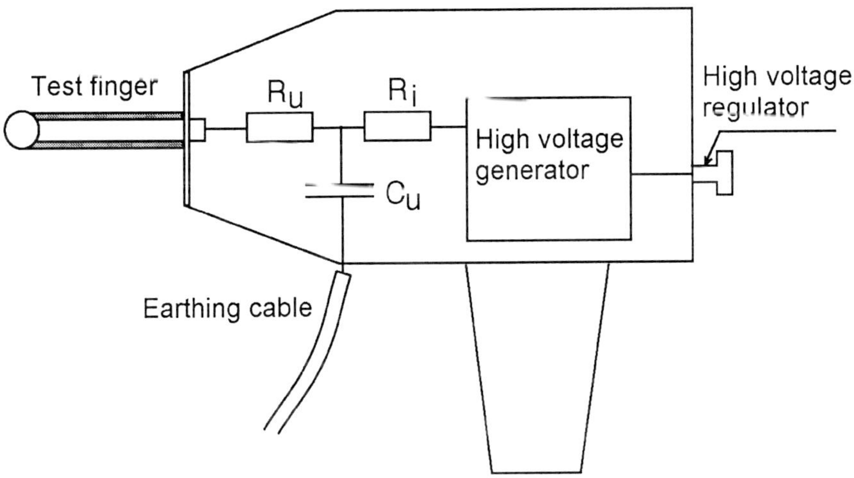

Fig. 106.
Principle of ESD-generators
R_i = Charging resistor (high resistibility)
C_u = Capacitor for energy storage
R_u = Discharge resistor

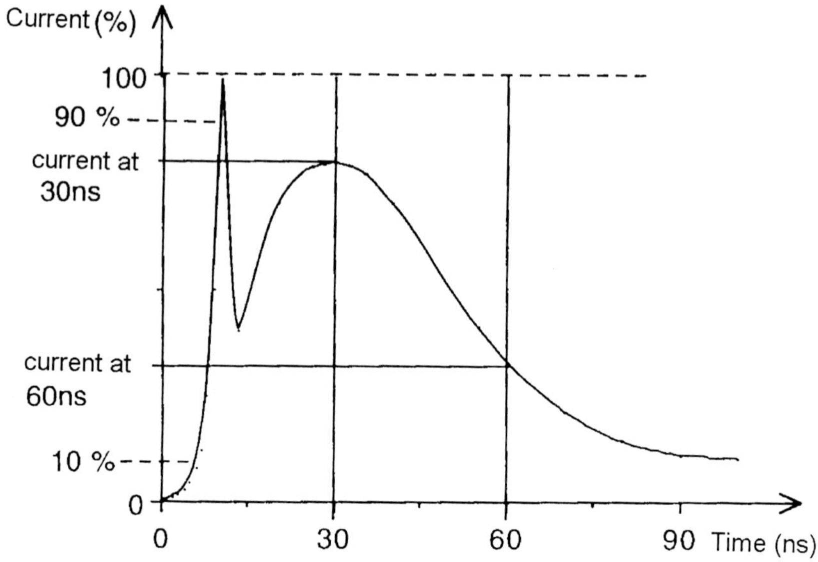

Fig. 107. Typical shape of the pulse from the ESD-generator

7.3.3
Bit-error measurements

The arrangement given in figure 109 is used to find the level for "Bit-error" and synchronization error, respectively. The antenna probe applicator for field strength measurements, is either most sensitive to E-fields or most sensitive to H-fields. The screen shall as in figure 104 have the distance 10 cm from the object to achieve a regular measurement. The field is measured at different distance from the screen. The composition of the probe is given in next figure.

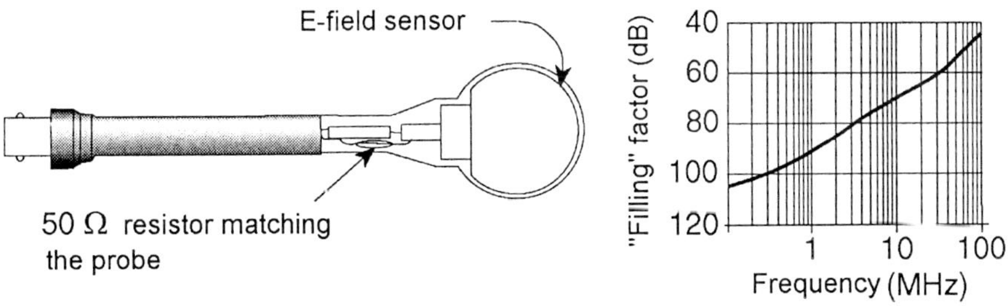

Fig. 108. Probe with a 3.6 cm "Sphere" for E-field having 2.3 GHz limit frequency and the characteristics of it

The so called antenna factor gives the ratio between induced field and the measured induced antenna voltage: PF = E/V (PF = Performance Factor) in dB. Other probes with loops are mainly sensitive to the magnetic field. The probes are calibrated in a TEM-cell as earlier described (figure 94).

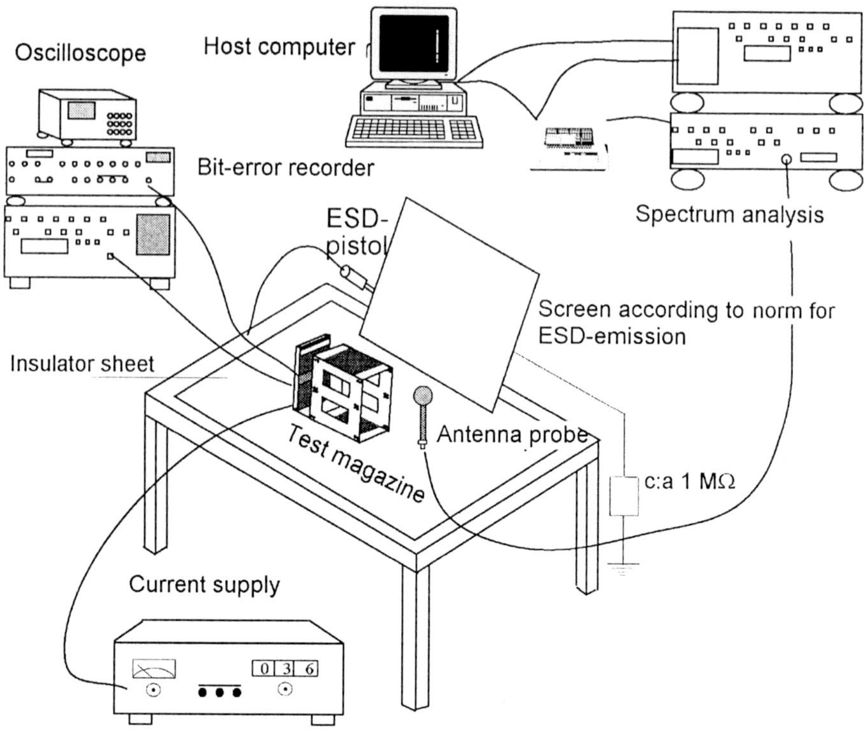

Fig. 109. Experimental arrangement for recording of bit-error caused by indirect ESD

7.3.4 Testing at PCB-level

Finally it can be mentioned that particular ESD-tests at PCB-level might be introduced applying the described ESD-simulator (pistol). See fig.110.

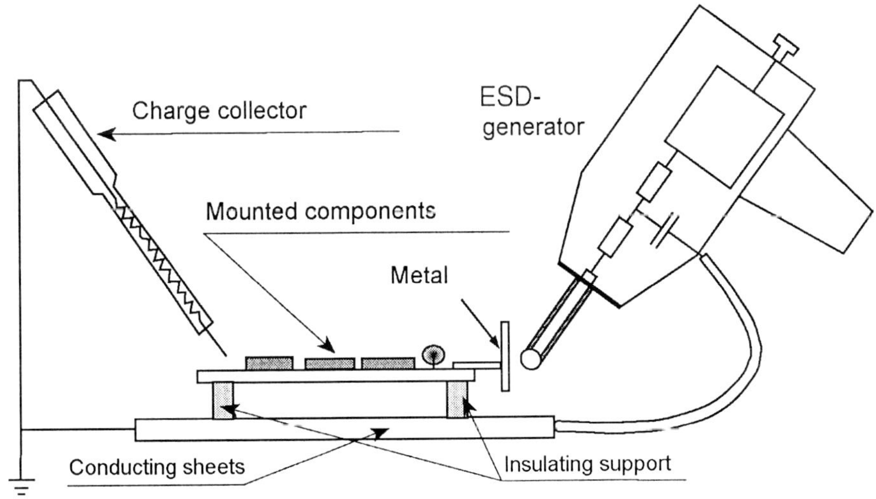

Fig. 110. Schematic picture of an earlier testing procedure suggested by Bellcore. First the charges are transferred from the pistol to the metal. Observe that the distance between pistol and the metal - is not critical! The charges are then transported to the conducting parts of the board (connection parts). A discharge can take place through a component to earth possibly destroying it. Between the "shots" the charges are collected with the charge collector. This is a very hard test

Bellcore made once a suggestion of the above testing method for PCB:s as they were worried about a lot of field failures due to ESD-damage.

About suggestions to protect PCB:s against ESD-damage can be mentioned the claim of mini-distance from component to the edge of the board. Another is to apply a guard-ring around the PCB. The most extreme is to have a "ring" around every component. This will be a question of cost. Committees within IEEE have been working on these problems.

There are some general rules for layout of PCB:s to protect against ESD. It is important to avoid large voltage loops for signals to earth or supply voltage to earth. Supply voltage or signal voltage shall have a return to earth through adjacent conductors. Earth plane or large copper sheets can act as perturbation free earth and screen.

8 Other types of perturbations in electronics

8.1 Noise

Noise is the most common perturbation in electronics. Taken in its most wide sense any disturbance or signal apt to hide the proper signal characterizes a noise phenomenon. It can be so called ripple from the current supply due to insufficient filtering. Noise is generated in microphones from vibrations of its components. Signals from outside can interfere with signals inside a system. Certain perturbation phenomena are able to be suppressed or totally eliminated, other are beyond control. The most well-known source of noise is, however, the thermal motion of the electrons in electronic components. This has a statistical character and appears both in active and passive components. Likewise the proper signal the noise is amplified!

Modern signal processing technique gives possibility to improve the signal to noise ratio (S/N). It is common to define a noise factor by:

$$F = \frac{(S/N)_{in}}{(S/N)_{out}}.$$

Other definitions exist.

Already in 1928 fundamental investigations on the statistical noise phenomenon were performed. Both an experimental and theoretical study were presented that year. The former achieved by J B Johnsson and published in Phys. Rev. Vol., 32 p. 97 July 1928. The other was fulfilled by H Nyquist and published in the same Journal and number, p. 110. The last author is well known for his "Nyquist criteria" concerning feedback from amplifiers. Both investigations led to the well known expression "the thermal noise voltage" ("Johnsson noise") represented by the square mean value for noise within a frequency range Δf. We define the "square-mean" of the noise voltage:

$$\overline{v_n^2} = 4kTR\Delta f \text{, where:}$$

T = absolute temperature
k = Boltzmann's constant
Δf = noise band width, depends only on the circuit
R = equivalent resistance of the circuit

The "noise" can be represented by a voltage source and a serial resistor.

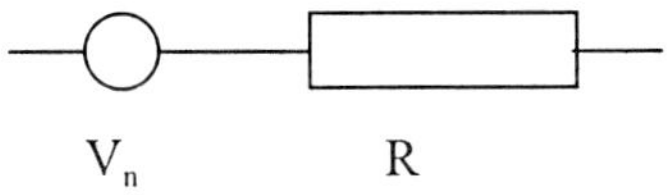

The noise current is corresponding to:

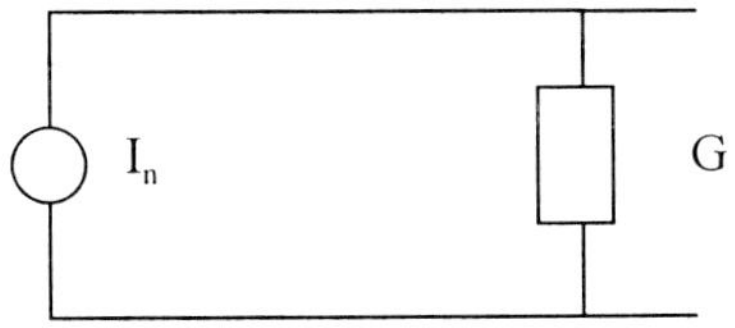

$\overline{i_n^2} = 4kTG\Delta f$;
G = conductance

Quantimechanically the following, generalized formula was derived:

$$\overline{v_n^2} = 4kT \int_{\Delta f} R(f)p(f)df$$

where $p(f) = \frac{hf}{kT}\left(e^{\frac{hf}{kT}} - 1\right)^{-1}$; h = Planck's constant

For frequencies up to 10^{11} Hz and room temperature hf/kT is << 1; by developing the exponential term in a series we get

$$p(f) = \frac{hf}{kT\left(1 + \frac{hf}{kT} + \frac{1}{2}\left(\frac{hf}{kT}\right)^2 + \Re_3 - 1\right)} = \frac{1}{1 + \frac{1}{2}\left(\frac{hf}{kT}\right) + \Re_2}$$

R_2 and R_3 are called rest terms
If hf/kT << 1; p(f) ≈ 1

This gives:

$$\overline{v_n^2} = 4kT \int_{\Delta f} R(f)df$$

Because of this "flat" spectrum the term "white" spectrum is used (compare with "noise" in optics). The limited band width Δf, depends only on the circuit not the noise source!

As electron tubes still are in use reference is made to the "noise formula" for a vacuum diode, also valid for "tunnel" diodes:

$$\overline{i_n^2} = 2qI_{dc}\Delta f$$

where q = charge of the electron $1.6 \cdot 10^{-19}$ As
I_{de} = the direct current part of the diode current

The formula is valid for frequencies $f \ll 1/\tau_t$. τ_t = the passing time of the electron through the diode $\approx 10^{-9}$ s. The noise is called "shot noise" and depends on variations of the cathode's electron emission. Similar formulas can be deducted for triodes and other electron tubes. If the diode is functioning in the "space charge-limited" region the variations are reduced by the space charge and the formula is changed to:

$$\overline{i_n^2} = 2qI_b r^2 \Delta f$$

I_b = the diode current at the operating point
r^2 = limitation factor due to the space charge, depends both on temperature and anode voltage

Similar calculations can be done for transistors. They are subject both to thermal noise and "shot noise". Typical for transistors is the 1/f-noise. This originates from surface effects and can be reduced by suitable treatment of the surfaces. The last mentioned noise is rather small and can in general be neglected compared with other noise effects. For an pnp-transistor the following noise sources are referred to:

Noise from fluctuations in the injection of "holes" from the emitter, that are to be collected at the collector.

Hole-electron recombination in the base region.

Thermally generated "holes" in the base region and injected to the emitter and collector regions. See fig. 111.

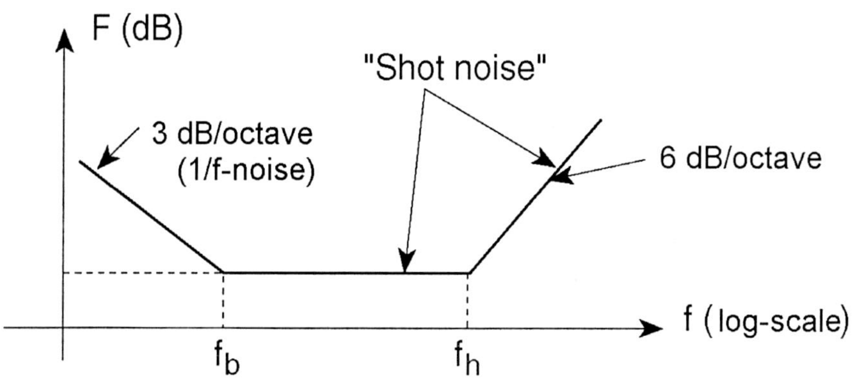

Fig. 111. Approximated transistor noise-frequency curve

For a more extensive account of noise phenomena the reader is referred to the literature on the subject.

A certain source of noise perturbation are so called dislocations treated in the science of solid state physics. Dislocations implies as the word says a "movement" (of atoms). The simplest type is a "point defect", an atom is displaced out of the site in the lattice and the potential function of that point is changed. Other types are "edge dislocation" and "skew dislocation". The following pictures are comprehensible descriptions of dislocations.

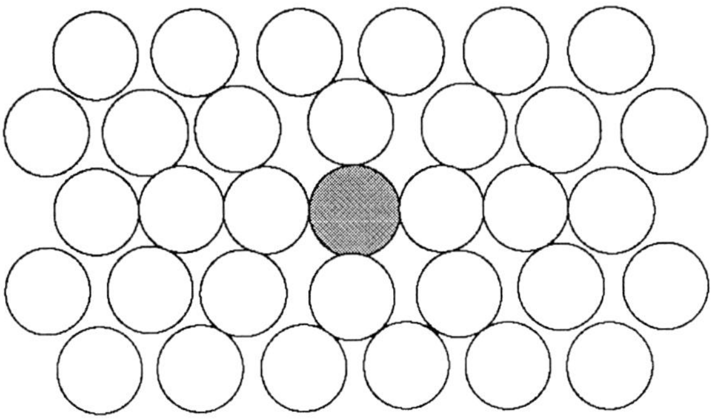

Fig. 112. Model of point defect. An extraneous atom disturbs the order of the crystal structure around the inclusion

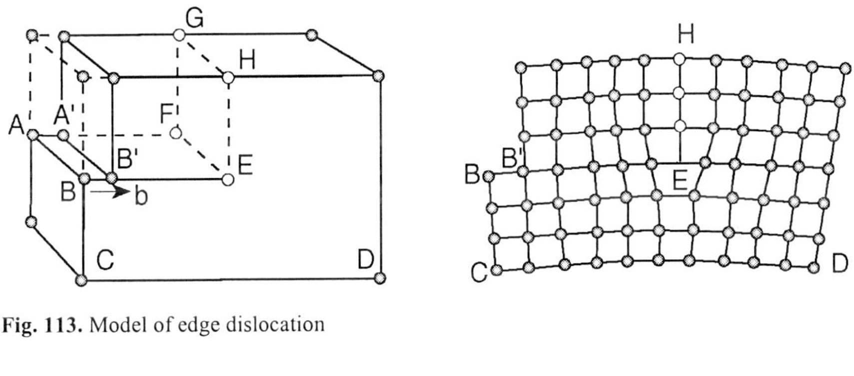

Fig. 113. Model of edge dislocation

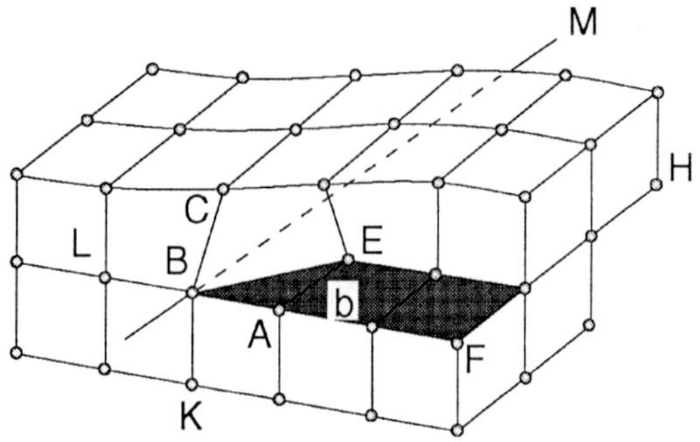

Fig. 114. Model of skew dislocation

The noise phenomenon comes from the generation-recombination (g-r) centra of the dislocation region. The relative direction between dislocation and current direction interferes. The space charge around the dislocation makes that the charge carriers are delayed somewhat changing the free mean-time τ_n. The current pulse is given by:

$$\int_{t_0}^{t_0+t_n} I dt = \frac{q}{l} \mu_n E \tau_n$$

where l= crystal length in the semiconductor
μ_n = electron mobility
E = electric field strength (V/l)

Thus the current pulse is proportional to the local electric field and the free mean-time of the charge carriers.

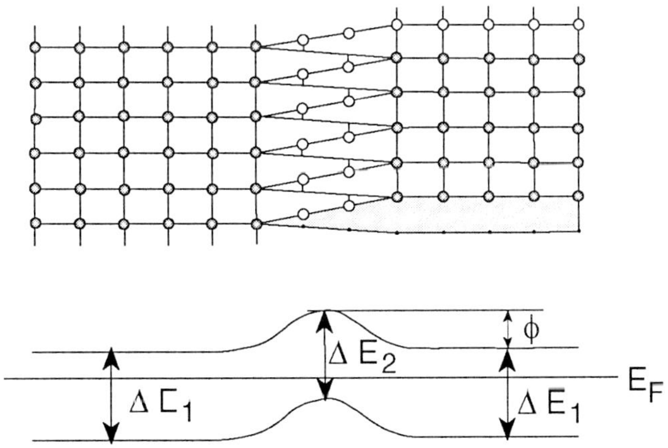

Fig. 115. Model of a perfect skew dislocation. From the scheme it is seen how the bandgap is influenced by the dislocation, that gives an impact on the current transport. E_F = Fermi level. ΔE_1 = bandgap before and behind the dislocation. ΔE_2 = bandgap at dislocation. $\Delta E_1 \neq \Delta E_2$

Dislocations acting as diffusion and recombination centra make τ_n shorter but contributes with more current pulses. The quadratic mean-value of the noise current will be:

$$\overline{i^2} \approx (q\mu_n E)^2 \text{ for a frequency f}$$

In the range f to f + Δf the following integral formula is obtained, if all Fourier components are used:

$$\overline{i^2} = \frac{1}{l}(q\mu_n E)^2 n_0 \tau_n C \int_f^{f+\Delta f} \frac{\sin^2\left(\omega\tau/2\right)}{\left(\omega\tau/2\right)^2} df$$

where C = lattice crossection

z = number of supposed current pulses

n_0 = mean-value of electron density $= z\dfrac{\tau_n}{C \cdot l}$

For the deduction of this formula see literature Sautter, 1956: Noise in semiconductors, published in "Progress in semiconductors", Gibson Kroger, Burger (editors), N.Y. 1956, Vol. 1, p.101. Compare also with fig. 116.

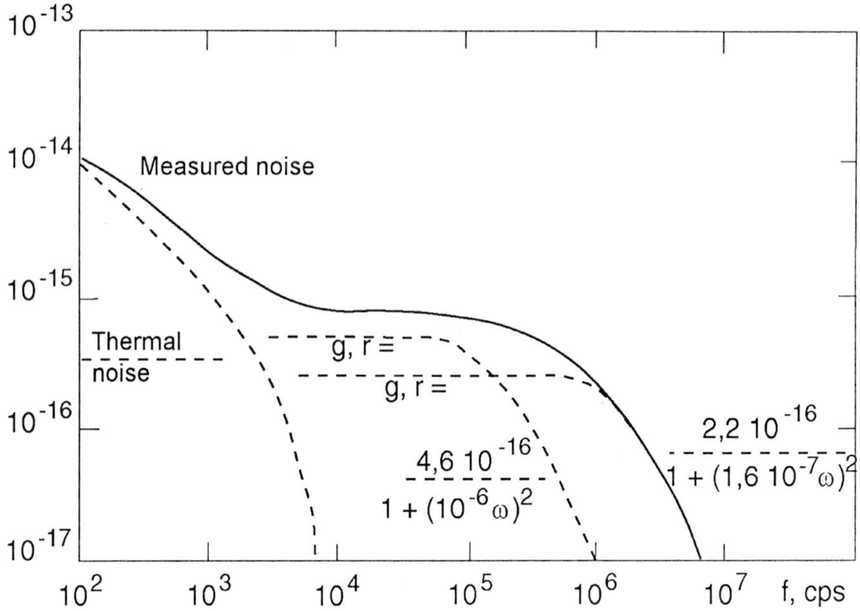

Fig. 116. Graphical resolution of total noise in 1/f- and g-r noise spectra at 106°C.

About noise from dislocations was also published by others (Jordan and Longini, 1967) as the fig. 116 for instance, J. Appl. Phys. Vol. 38, nr 2, 1967, p. 572.

Semiconductors with traps exchanging electrons with the conduction band and holes with the valance band get noise from these processes, too, see referred literature. Non-linearities, for instance in contacts, contribute also to noise generation.

8.2 Perturbation from transients

The diagram below gives roughly a comparison between different transients that can have an impact on electronics, directly or indirectly.

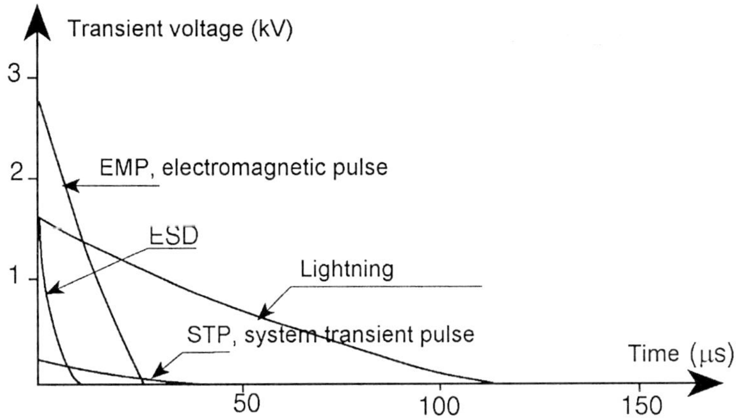

Fig. 117. Characteristic shape of different transients

The ESD-pulse is special because of its short duration. The origins of ESD-pulses were given before. Common transients from the mains are reduced by filters. Rectification of alternating currents generates ripple, that is also flattened by filters. Current supplies for electronics have filtering circuits.

By the external power lines transients can be introduced from a lightning for instance. Different performances of "transient protection" circuits are used to stop the transients. However, very fast, high level transients can yet reach attached electronic equipment.

The most extreme transient is NEP (Nuclear Electromagnetic Pulse), that is electromagnetic pulses radiated from a nuclear explosion (atomic bomb). This is gamma radiation of very high energy, that can have a catastrophic impact on electronics. Otherwise protected switching centrals could earlier (when the atomic bomb was a potential threat) be supplied at their inputs with a falling axe able to cut off all wires in a critical situation, and no dangerous transients would reach the switching equipment. Many theoretical studies were done about the impact on electronics from NEP and other radiation.

Semiconductors are sensitive to both high gamma doses and particle radiation as neutrons. This was studied on electronic equipments inside nuclear reactors.

Simulation programs for electronic circuits in a radiation environment usually include a special part for representing the influence of radiation in form of a voltage or current source. Electronic parts in satellites are exposed to cosmic radiation and so called radiation - hardened components are used. These are rather insensitive to latch-up. This phenomenon implies that parasitic transistors start working like a thyristor. They are initiated and cannot stop until the driving voltage is lowered to zero.

Defect free silicon oxide layers are resistant to radiation but irregularities of the structure become easily a centre for defect increase. On the reversal neutron irradiation can be used for doping of a component as the isotope induced might be the element wanted for doping.

8.3 EMC

8.3.1 Emission and immunity

The most common form of perturbation in telecommunication equipments depends on the sensitivity of electromagnetic radiation. The equipment itself can also have a leakage of electromagnetic waves disturbing other communications, a TV-broadcasting for instance. These phenomena represent immunity respectively emission properties of an equipment. For this a concept was introduced, EMC = Electro Magnetic Compatibility.

Immunity properties imply how an equipment or apparatus reacts and behaves when exposed by electromagnetic radiation. Emission properties imply consequently how electromagnetic radiation escapes from an equipment/ apparatus.

8.3.2 Design rules for EMC

The ever increasing use of electromagnetic waves for communication has extended the requirements of "compatibility" the last years. Definite limits are introduced. Design rules concerns mainly emission properties according to EN55022 at electronic circuits and systems, but to some extent also immunity.

Two methods are general available for optimization of the compatibility of electric equipments. One is to make an layout on the PCB:s of the circuits with smallest possible areas surrounded by conductors, i.e. smallest antenna function. It is also possible to array conductors with opposite current directions near each others, weakening the magnetic field! The other method is to screen off by conducting shields or enclosures. To get a sufficient result all holes and slits must as far as possible be shielded and movable parts (enclosure doors etc.) be supplied with packing strips of attenuating material.

Rules for circuit design are elaborated, even component design is optimized. Especially the encapsulation can cause different radiating effects. Minimization or compensation of even small inductance is desirable. The rise-time of pulses in logic circuitry has an influence on the creation of high frequencies, fig. 118.

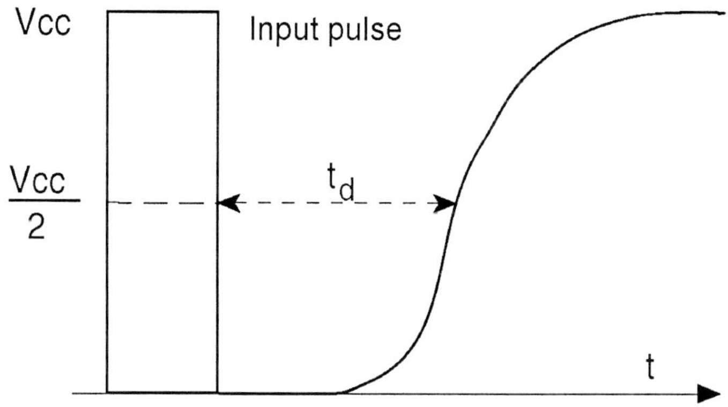

Fig. 118. Schematic picture showing how the delay-time is measured

Too "fast" pulses can increase the emission. Rise-time less than 0.5 ns should be avoided. Fast circuits giving desired short delays must be optimized as to delay/emission.

Some simple rules will be referred to. There are many handbooks for guidance about EMC. Other viewpoints than EMC must also be considered in the layout works, e.g. thermal requirements and ESD-hazards (breakdown etc.).

PCB:s should have at least 2 layers, multi-layertypes are even better.

A sufficient large earth plane is necessary.

Signal wires shall always be placed near an earth loop.

Avoid to mix analogue and digital circuit elements and common (floating) earth of them.

Avoid large "loops" on the board.

Use if possible large dimensions for earth conductor (width/length > 0.25).

A well defined impedance characteristic for fast logics, interfaces, rear-planes and "buses" should be given.

Couplings between PCB:s and to terminations shall be beyond magnetic fields from the board.

Signal wires, clock-currents, supply systems and earth systems shall not be laid around the PCB.

All logic should be physically restrained and decouple magnetic fields to earth plane.

Clock loops and critical signal wires shall be accompanied all the way by an earth loop. Use RC- and LC-filters for connecting a source with high HF- and RF-currents.

8.3.3 Testing of EMC-qualities

The electro-magnetic radiation is measured by antennas with performances described in chapter 7 for measurement of indirect ESD, i.e. biconical and log-periodic antennas.

Different norms were elaborated as to emission and immunity properties of electronic equipments. The normative requirements for emission and immunity are given by the EN-norms mentioned for "indirect ESD". This concerns especially what is called ITE = Information Technology Equipment. Very relevant are EN 55022 (Radio perturbation from ITE, corresponds to so called CISPR-norm number 22) and EN 55024 (Immunity characteristics of ITE, corresponds to CISPR24). A lot of other norms for electromagnetic radiation exist.

The bit-error measurement described in chapter 7 is an example of immunity testing. To test the immunity against continuous radiation the screen in the example can be replaced by a radiating biconical or log-periodic antenna. Corresponding international standards for immunity are IEC 801-3, IEC 801-4 and IEC 801-5.

Alternatively to EMC the term EMI = Electro Magnetic Interference is used, including all kind of electro-magnetic perturbations of equipments and apparatures.

ITE is classified in two groups. A and B. For class A a fixed limit of perturbation higher than for class B is determined. Because of this class A equipments can be subjected to restriction in certain countries to selling and use. For class B a lower perturbation level than for class A is fixed (a more severe demand than for A). No restrictions as to selling or, in general, use in this case. The following tables summarises the valid limits for class A and B.

Frequency range (MHz)	Limits [dB (μV)]	
	Quasi-pic	Mean
0.15 - 0.50	79	66
0.50 - 30	73	60

Table 4. Perturbation levels according to CISPR, class A, for frequencies: 0.15 - 30 MHz

Frequency range (MHz)	Limits [dB (μV)]	
	Quasi-pic	Mean
0.15 - 0.50	66-56	56-46
0.50 - 5	56	46
5 - 30	60	50

Table 5. Perturbation levels according to CISPR, class B, for frequencies 0.15 - 30 MHz

Frequency range (MHz)	Quasi-pic - limits [dB (μV)]
30 - 230	30
230 - 1000	37

Table 6. Perturbation levels according to CISPR, class A, in frequency range: 30 - 1000 MHz measuring distance 30 m.
Perturbation levels according to CISPR, class B, in frequency range: 30 - 1000 MHz measuring distance 10 m.

Finally a comparison is shown between American (FCC) and universal (CISPR) limits of class A and B at different measuring distances, fig. 118.

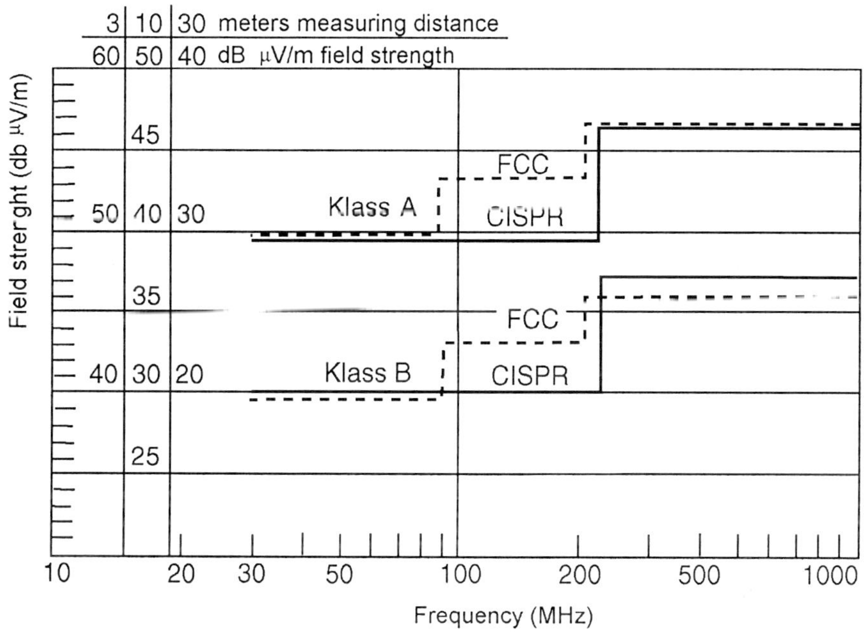

Fig. 119. FCC, Part 15, Subpart J and CISPR, Publ. 22. Emission limit measured at distances 3, 10 and 30 meter.

9 The impact of electric fields and electromagnetic waves on the organism

9.1 Computer screens, TV-sets etc.

Since the very beginning of creation we are surrounded by different forms of radiation. Sunshine and cosmic radiation. The former is of electromagnetic character, while the latter also contains particles of very high energy. Radioactive radiation emanates partly from the earth surface. This often contains uranium and thorium with their daughter products (as radium with the gas radon). It is anticipated that the ions in the air are created mostly by radioactive irradiation from the earth.

We can assume that we are adapted to these natural species of radiation. People have, in itself, different sensitivity against sunshine (ultra violet light, particularly can hurt human tissue). The earth has also a pertinent magnetic field. But what would happen, if the intensity for some of these radiations increased essentially above its normal value?

Because of the human eagerness to investigate nature other types of radiation have been released. The radioactivity implies a classical example of this, as well as the X-rays. The radiation being invisible people did not at first realise the health hazards involved. But by experience they soon learned to take precautions, i.e. protect themselves against the radiation.

Cancer is as known a common result of an overdose of irradiation, though, curiously enough, limited doses X-rays or γ-radiation can inhibit cancer growth, even kill the cancer cells. X-rays and γ-radiation are electromagnetic character, but having very high energy compared to light and radio waves.

Laser radiation, a type of light, has a unique position because of its coherence (phase invariance) and can have a strong effect on materials and tissue.

There are different types of radioactive radiation, particle (α^+ or β^-) - and γ-radiation. A nuclear disintegration by alpha- or beta emission is often followed by a gamma photon (photon = quantum = the unit energy of the radiation, proportional to the frequency of the gamma radiation).

Because of the increased use of electromagnetic radiation (IT for instance), a natural question is what influence it might have on the organism. This concerns also static electric and magnetic fields. The increased use of picture screens (computers and TV-sets) has made the question rather important, because these generate both electrostatic and electromagnetic fields. People, who more or less continuously at their works are seated in front of such screens have complained about indisposition or skin irritation. But no permanent symptoms.

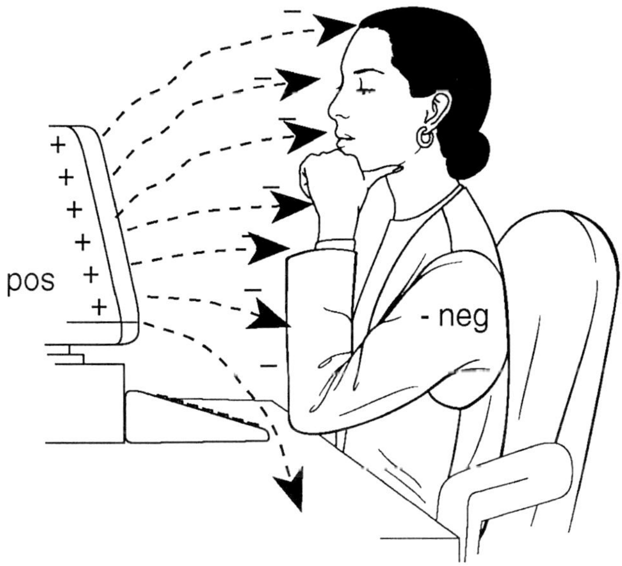

Fig. 120. Computer screens can create electrostatic fields when bombarded by electrons and optic radiation (light and heat). Low frequency (50 Hz) and electromagnetic high frequency fields are present. Out of these only the electrostatic field has proven to give effects on the organism. Most oftenly the skin is hurt. But it can be restored by medical treatment.
The inconvenient situation with a strong field from the computer, resulting in transport of positive ions, according to the picture, towards the face of the operator depositing accompanying dust particles onto it, can be avoided by placing a bipolar ionizer on the top of the computer. This will neutralize the screen. Alternatively an earthed screen with a conducting, transparent deposition is placed in front of the computer screen.Another solution to avoid electronic fields from screens is to use screens of liquid crystals.

Many investigations on scientific basis were performed to find the causes of the effects. In these all factors that could influence the results had to be considered. Followingly, except physical parameters also psycho-social aspects as stress situations are important factors. Chemico-medical contamination of the air can have similar effects.

Typical is that from investigations presented trends of influence on the organism by electromagnetic fields are found, but it is difficult to find a strong correlations to the electric environment. Some results will be accounted for. In general it is trade unions and the care centers of enterprises that take the initiative in this kind of research. The investigations are achieved at the working places but the research is prosecuted at universities, hospitals and institutes.

A Swedish investigation of 800 persons who had been working more than 20 h/week at computer screen for 5 years, proved that inconveniences were felt twice more often than for a control group working without computer screens. In another investigation 83 samples of tissue were taken to state the presence of permanent skin injuries after exposion from a computer screen. No permanent injuries could be found. A third investigation using a population of 430 000 governmental employees gave the following results as to influence of computer screen.

Number	Category	Symptoms
1150	1	Minor skin irritation, when working at computer screens. Recovered next day.
350	2	Skin irritations remaining for a longer time after work at a picture screen.
90	3	General "electric hypersensitivity" reducing working ability. Requires medical care, sick-leave and change of work.
60	4	Serious inconveniences also from electrical installations at home. Essential limitations in working and life conditions.

Among those interviewed in the above investigation there is with a high probability a "grey zone" of people who from different reasons do not want to reveal their problems with working conditions.

A fair estimate of the part of people working at computer screens and experiencing a pronounced inconvenience is about 0.5 to 1.0 percent (this has however not been proven). Most of the irritation are also of a mild character.

Evidently not only picture screens can cause inconveniences with people. Other electrical equipment can be annoying as luminous tubes and fans. This has been confirmed by an extensive investigation at an electronic company. Most of them who were complaining felt a weak skin irritation. A general result of the investigation was that not only picture screens as mentioned above was the origin of the discomfort.

Some results have stress situations as the main factor for inconveniences.

To get a possible significant correlation with the electric field from screens and irritations by the operators a Swedish test used two groups of people working at picture screens with different electromagnetic environment. The electric field strength from screens for one group was 30 V/m for the other 10 V/m. People exposed by the stronger field felt skin irritations three times more often than those with the weaker field! Radio amateurs can feel inconvenience when operating their radio equipments.

In USA they found that radiation of low frequency from so called video terminals (VDT) can have an impact on the human cells. If this also is a health hazard was not able to conclude due to few data. Not only the screen of a computer contributes to the electrical environment. The transformer, generally at the back of the computer, generates a field that can reach people behind or beside the computer, even behind the wall of the next room! The situation with missing legal prescription, for instance for use of VDT:s, was criticized in USA. The computer manufacturer pretend they follow given regulations, but these are not sufficient.

It was suggested before to use ionizers that neutralizes charges. But ions themselves could have some negative influence. This was investigated and discussed very much. It is often pretended that negative ions should have a refreshing impact on the organism, while positive ions promote tiredness, even depression. But there is no uncontradicted evidence for this. Many investigations were presented in the literature. As what is true for electromagnetic radiation the formation of the hormone melatonine is influenced, causing organic effects.

Due to the experience gained, situations with electric hypersensitive people can now be handled in a rational way taking also psycho-social factors into consideration. Electric fields are eliminated by screening or earth connection. An unearthed electric metal lamp can give a stronger electric field than a computer! Clothes of synthetic material are generators of strong electrostatic fields! If an unearthed cable to a table lamp is lying on a metal sheet, this will act as an antenna with a strong electric field as result!

Psychosomatic relations are sometimes formed in connection with sick symptoms. Still deeper psychological factors could be active. Freud would probably have called upon the "super ego". Special precaution must be taken for people in delicate situations as pregnant women.

9.2 Electromagnetic radiation fields

Another problem area where the opinions are very divergent is the electromagnetic radiation from high voltage power lines. Extensive investigations were made both in USA and Europe.

One result should be that cows which have been grazing for a long time near power lines give less milk than normally. There is no evidence for the gene plasm to have been changed in any case. Mutation of genes can also be initiated by certain chemical substances.

9.2.1 Electric fields

Voltage between long wires and a plane, (earth as an example) give rise to an electric field, E field, measured in kV/m. Unlike the case with point sources (Chap 7) the field shape can simply be given as the E-field between two parallel plates, figure 121.

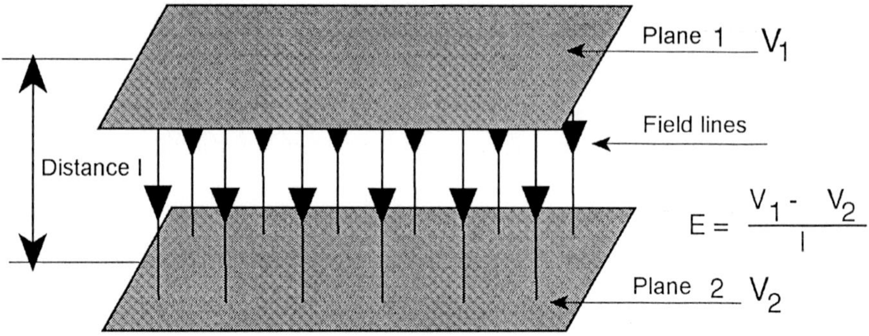

Fig. 121. The electrical field between two planes. If the voltage between them is 2000 V and the distance L= 1 m the field strength will be 2 kV/m

In a real case several factors will change the field. The field lines are perpendicular to the conductor surface. Thus the field is distributed over a surface larger than that direct beneath the power lines, figure 122.

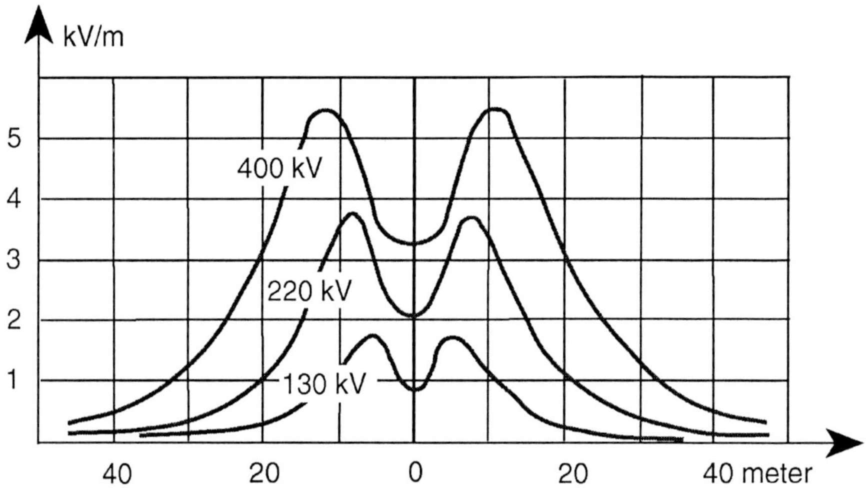

Fig. 122. Electric field strength measured beneath power lines at 1.8 m above the earth plane

When measuring the field-strength the height of a person is approximated to 1.8 m (see figure). If somebody is carrying a long object the field around him will be changed.

9.2.2 Magnetic fields

Currents in the power lines are associated with a magnetic field, B-field, the unit of which is tesla (T). The dimension of tesla is Vs/m^2. At a point with distance r from a straight long conductor, where r << l (l = length of conductor), B will be, according to next figure:

$$B = \frac{\mu_0 \cdot I}{2\pi r} \qquad I = \text{current}$$

$$B \approx 2 \cdot 10^{-7} \cdot \frac{I}{r} (T)$$

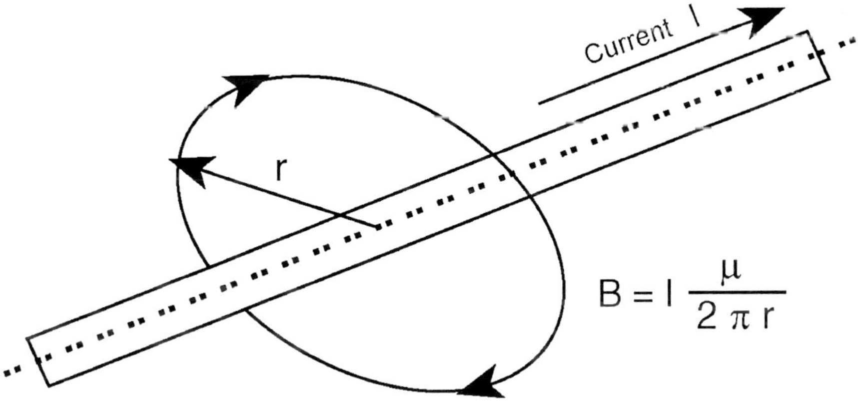

Fig. 123. Magnetic flux density (B-field) around an extended conductor. A point at a distance 1 m from the conductor with current I = 500 A is exposed to a B-field of 10^{-4} T or 100 μT.

Magnetic flux density (B-field) around an extended conductor is strongest near the conductor and decays with the distance from it (Figure 124). The formula for "B" is correct immediately under the conductors. But when the field is entering the earth the permeability will change, i.e. μ_0 is replaced by $\mu = \mu_r \cdot \mu_0$, where μ_r, the relative permeability is not 1 as in air.

The E-field around power lines are much stronger than elsewhere (e.g. inside houses). The B-field, however, is even near power lines equivalent or weaker than what is normally measured inside a house. Domestic accessories, razors and electric alarms give B-fields about 10-500 mT.

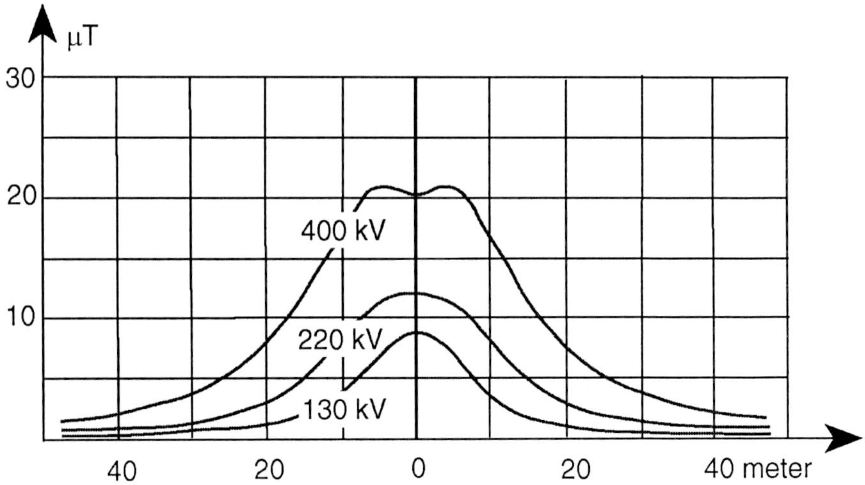

Fig. 124. Magnetic field strength from power lines, measured at point 1.8 above the earth plane.

9.2.3 Effects on people and animals

The probable first report about possible effects on human beings from electromagnetic fields came from the Soviet Union in the middle of the 60-ties. The investigation concerns workers in an interlocking-plant with E-fields as high as 25 kV/m. They suffered from headache, tiredness, sleeping difficulties etc. It was also anticipated that stay of long duration in the strong field might hurt the central nerve system. This report was published in Paris 1972 and attracted together with a Spanish report great attention. The reports initiated other investigations round the world. None of these investigations gave evidence for effects on people exposed to E-fields.

Even if these first investigations about E-fields gave negative answers, results from B-field of low frequency indicated leukemia hazard especially for children.

In the early investigations it was also not possible to confirm influence on animals from electromagnetic radiation. For instance cattle staying long times near power lines of 365 to 765 kV were studied in different countries. Milk production or fertility remained undisturbed.

But this has not discouraged people to continue the search for a possible connection between electromagnetic radiation and sicknesses in living species.

9.2.4
Recent trends in the research

Many new results about effects of electromagnetic radiation are the organism have been accounted for and new approaches are applied. The clue is the interaction mechanism electric fields-organism. Magnetite (Fe_3O_4), a ferromagnetic mineral has been reported to be present as trace elements of varying amounts in different tissues as brain and tumors (Kirschvink J L et al, 1992). Mechanical vibrations of magnetite particles in an alternating 50 or 60 Hz electromagnetic field are suggested to bc the damaging process, but this is only a hypothese. Birds might orient by magnetite particle in earth magnetic fields.

There are other medico-chemical approaches, too complex to be within the scope of this account. We have to wait for their evaluation. Especially the appearance of tumors is of large interest. A great lot of new, valuable information will probably be released in the near future.

10 Simulation methods

10.1 Mathematical approach

The temperature is a crucial factor in the damaging of components, exposed to ESD-transients. A way to study the influence of the temperature is to simulate mathematically the behavior of components, when stressed by electrical transients.

At first analytic formulations were tested for interpretation of the heat diffusion in the component. Some results were gained in that way by using the heat-transport equation. The most well known issue is the Wunsch-Bell model (Wunsch and Bell,1968). According to this model the power, P, needed to cause a damage is inversely proportional to the square-root of the pulse-width (square pulses), t_f :

$$P = \frac{1}{\sqrt{t_f}}$$

In this interval also ESD-transients are found, fig. 125.

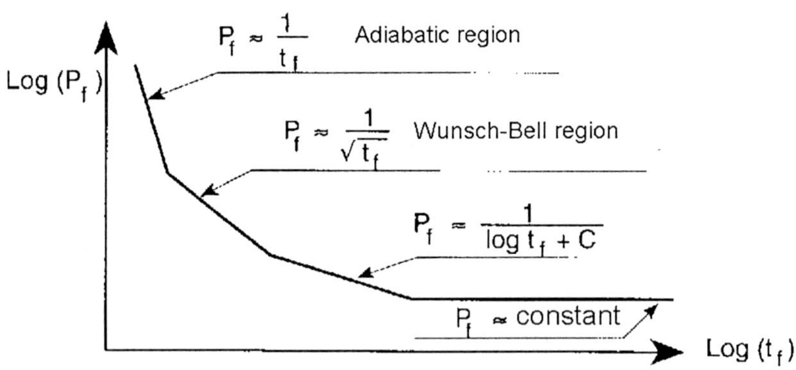

Fig. 125. Power density at an pn-junction to cause damage as a function of applied pulse width of the transient, including the Wunsch-Bell model

These approaches were done in one dimension with constant parameters. It has been proved, that they hold even if the temperature variations of the parameters are introduced (Ash, 1983). Some analytical expressions for the dependence of parameters on the temperature were also deduced. A global solution for the transient behavior of semiconductors, including temperature, was the next step, but still in one dimension (Orvis et al, 1983).

The electrical behavior of a semiconductor can be described by the continuous electron (e) and hole (h) equations together with the Poisson equation:

$$\frac{\partial n}{\partial t} = G - U + \frac{1}{q}\nabla J_n$$

$$\frac{\partial p}{\partial t} = G - U - \frac{1}{q}\nabla J_p$$

$$\nabla^2 \varphi = \frac{-q(p - n + N)}{\varepsilon}$$

where $J_n = qn\mu_n E + qD_n \nabla_n$

$$J_p = qn\mu_p E - qD_p \nabla_p$$

J_n and J_p are the current density equations for electrons and holes, n and p being the electron and hole concentrations. G (|E|, T, n, p) it the avalanche generation term, U (T, n, p) the recombination term. φ is the electrostatic potential, q electric charge and ε the dielectric constant. N is the excess of positive doping concentration.

μ_n (|E|, T, N) and μ_p (|E|, T, N) is electron respectively hole mobilities and E the electric field

D_n (|E|, T, N) and D_p (|E|, T, N) are finally electron and hole diffusion coefficients.

To include temperature, the solution of these equations is coupled to the heat transport equation:

$$\rho c \frac{\partial T}{\partial t} = \nabla(K \nabla T) + |JE|$$

Where:
- T = absolute temperature
- ρ = density
- c = heat capacity
- K = thermal conductivity

Many of the parameters in the equations above are temperature dependent, which increases the influence of temperature and the non-linearity of the system.

They form a set of coupled, non-linear differential equations, that are difficult to solve, why numerical methods must be used.

To simulate the occurrence of failure in semiconductors due to the temperature rise, a 2- or, better, 3-dimensional simulation model should be used. The solution of the corresponding equation system will encounter convergence problems. But now exist, however, several efficient computer programs, that can be applied. A failure criteria can be the temperature reaching the melting point of the semiconductor material. Silicon is melting at about 1 410°C.

Another criteria could be the exceeding of the temperature, at which the eutectic composition of aluminium and silicon is melting, 577 °C.

Proper boundary conditions as well as initial conditions are important for mathematical treatment of physical processes.

10.2 Simulation of transients across an pn-junction

By way of illustration simulation in one dimension of an pn-junction is given.

The energy of a pulse dissipated inside a semiconductor will develop heat and raise the temperature. This rise is dependent on the thermal conduction properties of the component. Effective heat transport will slow down the temperature rise. Due to the extended numerical work with strong field gradients that appear at a certain point of time causing convergence problems, the simulation is restricted to the influence of the pulse front. As simulation object the pn-junction shown in the next figure was applied.

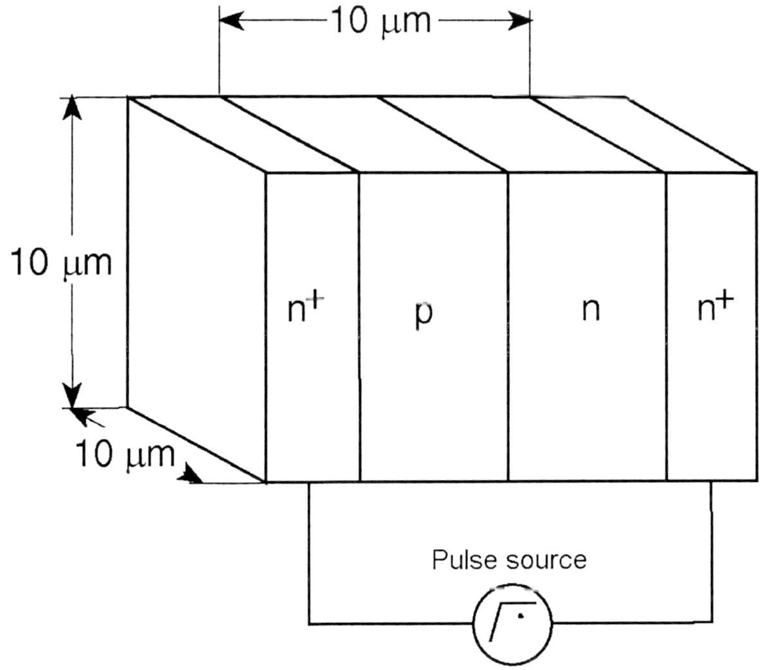

Fig. 126. Used model for pn-junction

The transient is applied in reverse direction, fig. 127 it becomes constant after about 1.3 ns at the level 220 V. Another boundary condition is the temperature being fixed to 300°K at the front side (0 μm) and back side (10 μm) of the transition region, fig. 126.

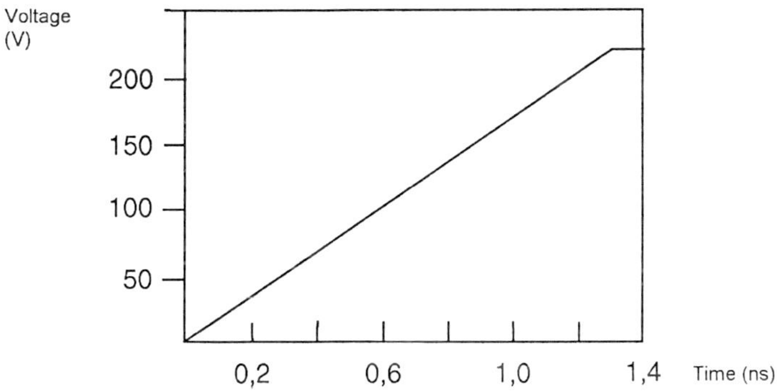

Fig. 127. Simulation pulse as function of time

10.2.1 Results

At the starting point (t = 0, V = 0) the doping profile is symmetric for both p- and n-regions. After 0.15 ns a depletion area is formed and the potential is displaced, fig. 128.

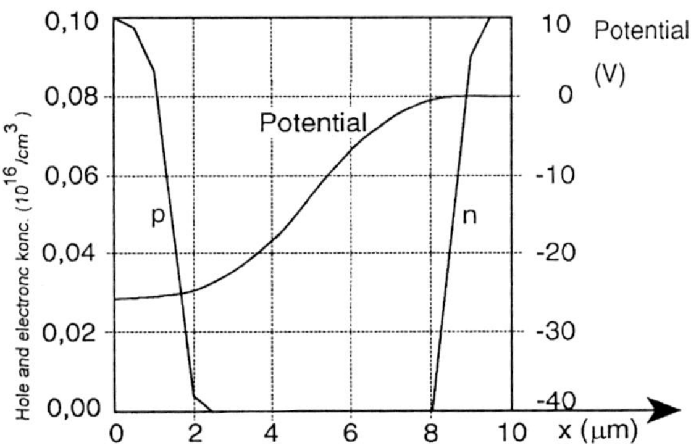

Fig. 128. Doping profiles of holes and electron concentration according to simulation and the potential at time t = 0.15 ns. A depletion region is formed

Only after 1.275 ns there is a substantial change, next figure.

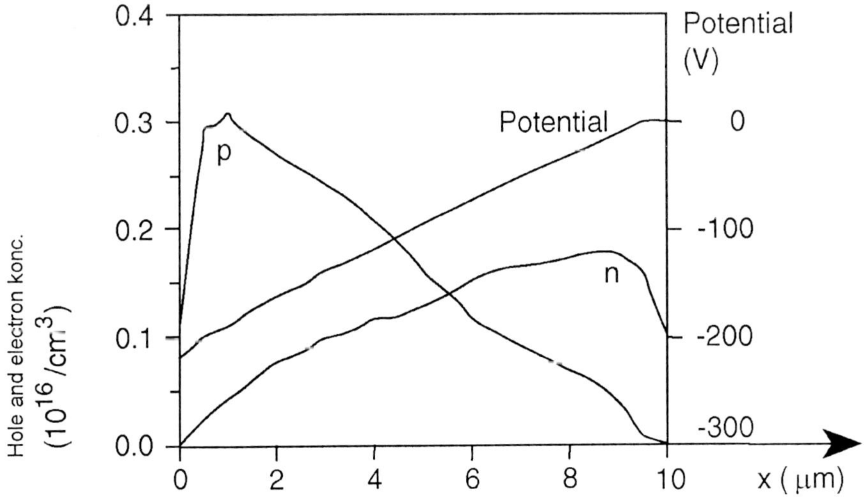

Fig. 129. Doping profiles and potential after 1.275 ns.

At 1.3 ns the potential distribution is rectitinear, see fig. 130. Hole and electron concentration are equal across the junction. The slope of the potential is different at the beginning and end of the junction.

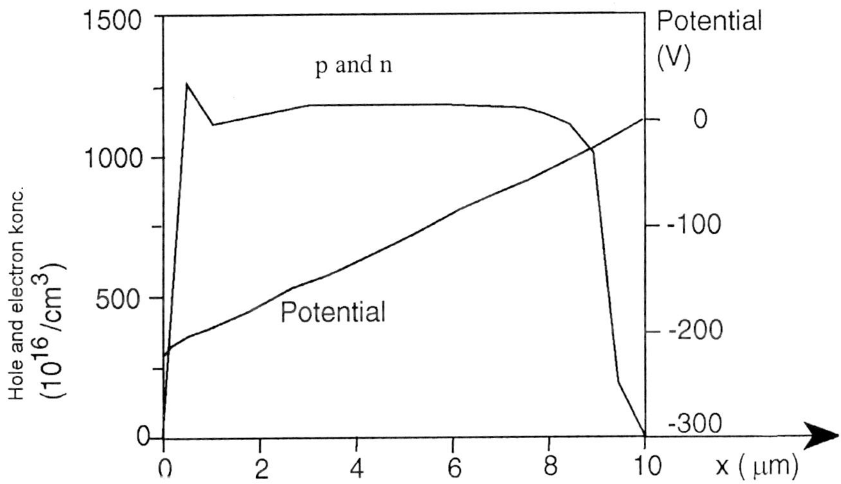

Fig. 130. Simulated doping profiles and potential after 1.3 ns

Most interesting is that the temperature now is increased substantially, se next figure.

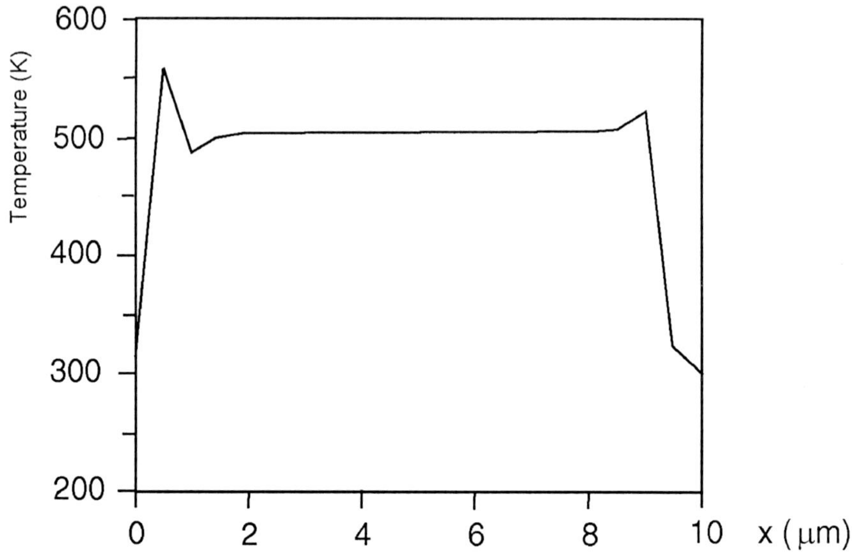

Fig. 131a. The temperature distribution after 1.3 ns

The results described correspond to the boundary conditions chosen, i.e. constant temperature at the input and output, of the pn-region (cooling). Other approaches with improved cooling at some point can be applied. Thus it is possible to compare by simulation different performances of pn-junctions as to geometry, heat conduction and doping profiles, different pulse rise-time and final level of applied transient.

Simulation using an ESD-transient, that decays rapidly, should give results showing less stress than in the given example. The results above agree with earlier results presented (Orvis et al, 1983).

Using a pulse with a shorter rise time for the simulation will give similar results but the time scale is shortened. A 5 times shorter rise time gave the results obtained at 1.3 ns before already at 0.2814 ns. But the results were identical.

10.3 2D and 3D simulations

Only qualitative results can be obtained by one-dimensional simulations of ESD events. 2D or better 3D analysis is necessary for a physically proper simulation.

The following figure shows 2D simulation of a bipolar component. The results were in good agreement with experimented results performed with HBM-ESD pulses using 500 pF capacitors and 100 ohm resistor (Maede and Wade, 1989).

Other parameters than temperature like voltage, current, electric field etc can be obtained from the simulation as function of time or space coordinats for instance. Hot spots are easy to detect.

Now there are programs available for 3D transient simulation. To improve the efficiency of the programs parallel processor techique is applied. ATLA-ATHENA and Spice are example of this kind of programs. As mathematical tools the finite element and finite difference methods are used. But other possibilities exist.

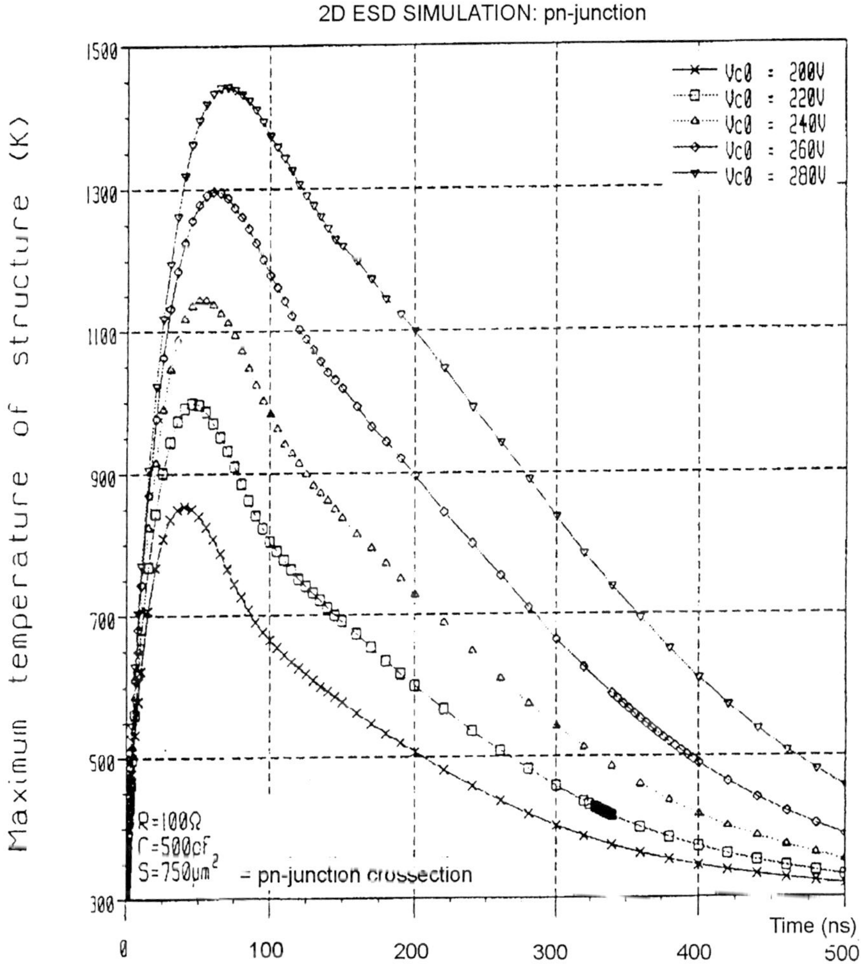

Fig 131b. Experimental results (Maeda and Wada, 1989) were verified by 2D-simulation. Compare also with fig. 131a.

11 Protection methods - Antistatic materials

11.1 General

In the sixties MOS-components were regarded as very ESD-sensitive. But when new component technologies were introduced, especially IC:s, it was realised that other circuit elements could be sensitive to ESD. The narrow conductors, the small distances between circuits and the thin oxide layers increase the sensitivity. Now all kind of electronic components even passive, are considered sensitive to ESD. This has started activities to avoid or preferable eliminate ESD hazards. A general measure in that direction is to increase the relative humidity (RH-value). By that otherwise insulating surfaces are rendered slightly conductive, thus counteracting charge formation. Alternatively ionizers can be used producing ions that neutralize charge of opposite sign, a method often used in clean rooms. The aim is to create an ESD-safe environment. But these actions are often not enough. This chapter will describe all protection methods. Another early measure was to supply electronic circuits with special protection circuits, incorporated in the circuitry. The function of them is to reduce voltage and currents of transients to a harmless level and dissipate energy developed by them.

11.2 Protection circuits in components

Input of IC-components are known to be sensitive to ESD. Outputs are also sensitive but often less than the inputs. That is why only the inputs were protected at first. An important fact is, as proved by tests, that a stressed input can remain unchanged while other in- or outputs get damaged. In modern circuits also the outputs are protected. The first protection circuits were designed to withstand HBM-stresses. But the increased use of automization in mounting and testing made protection against CDM- and MM-stresses necessary.

The basic principle of protection is to protect inputs by current limiting resistors, and voltage reducing elements, e.g. diodes. Fig 133-135.

Following figures depict some of the most common protection circuits and the development in their design. Modern LSI- and VLSI-processes must include steps for incorporation of protection circuits. But interference with other requirements of the circuitry must be avoided and the "latch-up" immunity conserved.

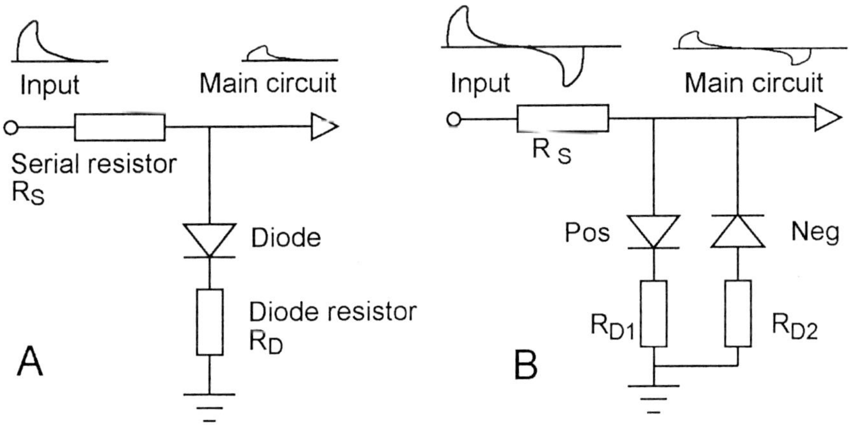

Fig. 132. A. Unipolar (One way) protection. B. Bipolar (Two ways) protection.
The parasitic resistors of the diodes, R_D, must not be too large or the protection efficiency is decreased

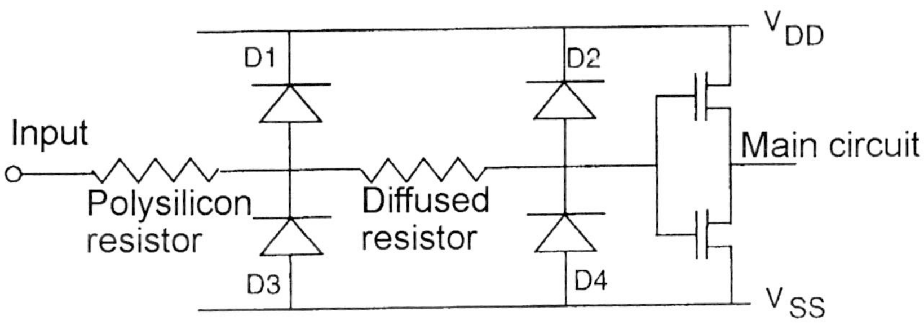

Fig. 133. CMOS-protection. The are led, dependent on polarity, to either the V_{DD} or V_{SS} connections by the diodes D1 and D3

The polysilicon resistor in fig. normally 200-300 ohms achieves an isolation to the diffused junctions and serves as a current reducing element and latch-up sensor.

The diffused resistor is limiting the current to the diodes, D2 and D4, and reduces the transient voltage at the input of the main circuit to a safe level.

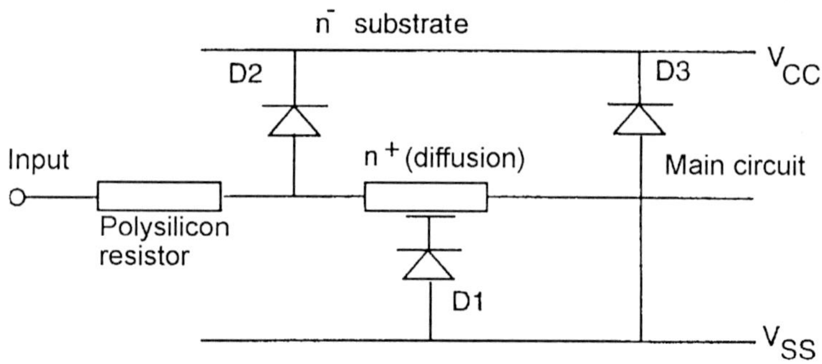

Fig. 134. HCMOS-protection

In figure.135 a n-doped substrate with a p-doped well is used. The diodes D1 and D2 are performed in the p-well. D1 reduces positive and D2 negative transients. The parasitic well of the substrate-diode D3 reduces transient voltages at the connector. The polysilicon resistor is limiting the current. It has a relative large effective surface (about 180 x 20 μm) to overcome the thermal problems associated with this type of resistors.

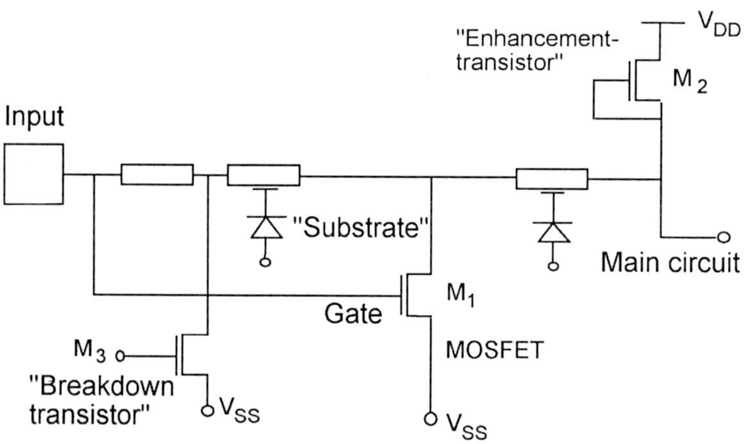

Fig. 135. VLSI-protection circuit. A "combined" protection circuit for fast VLSI-circuits. Protection is given against "high", "medium high" and "low" transient levels

In the VLSI-protection circuitry above fig 135 M_1 provides protection against fast positive high level ESD-pulses, because the gate is connected to the input. An ESD pulse opens M_1 to give a low resistance path to V_{SS}. The threshold voltage of

M_1 is 20-30 V, what gives no protection against low level pulses. The same is the situation with M_3 that gets conductive at 20-30 V.

As a complement there is a field-effect transistor, M_2, with thin oxide, activated at a voltage level less 1 V. It gives protection in the range 10-30 V. M_2 operates in the "saturated mode". Protection against negative pulses is provided by the thin oxide transistor formed from the diffused resistor that introduces a forward bias to the substrate.

Additional protection is obtained by parasitic NPN-bipolar transistors with reversed biased collector-base junctions (source-substrate of M_1 and M_3). Forward-biased emitter-base junction (drain-substrate of M_1 and M_3) give these a forward active mode and provides another low resistance path to V_{SS}. Next fig. shows the characteristics of thin oxide transistors.

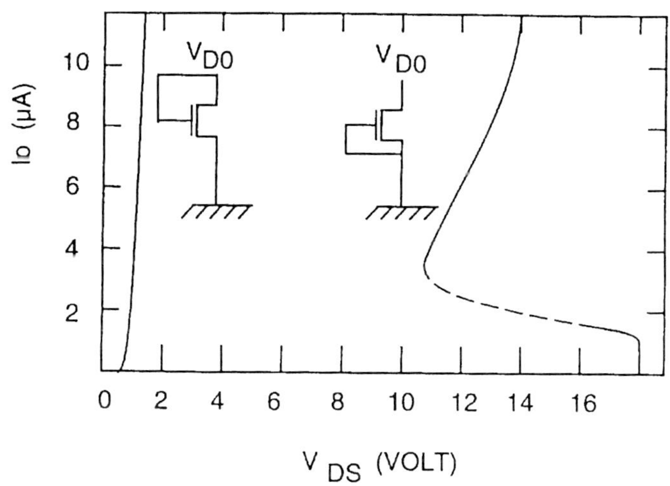

Fig. 136. A comparison between the operational modes of a thin oxide transistor

Delays introduced by the described protection circuit was tested by simulation and found to be of the order of 3 ns at a load of 0.8 pF. This circuitry is described in "Proc. of 1984 Custom Int. Circuits Conference IEEE N.Y. USA" under the title "Optimized ESD protection circuits for high-speed MOS/VLSI" by E. Fujishin et al.

The following protection circuitry is often used, figure 137.

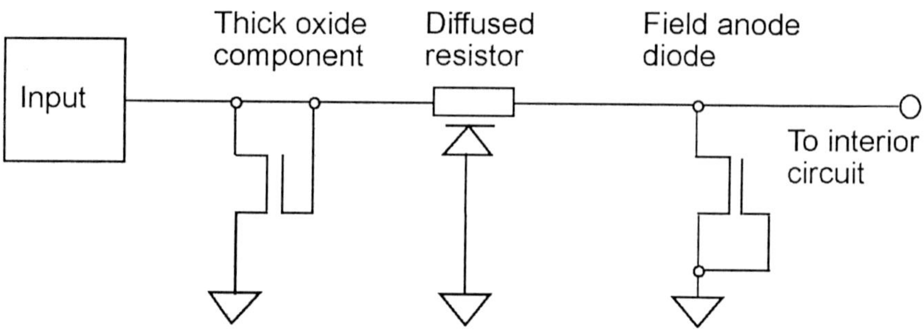

Fig. 137. Scheme of a modern input protection circuit

The following circuit, fig 138, is intended to protect against CDM-pulses, too.

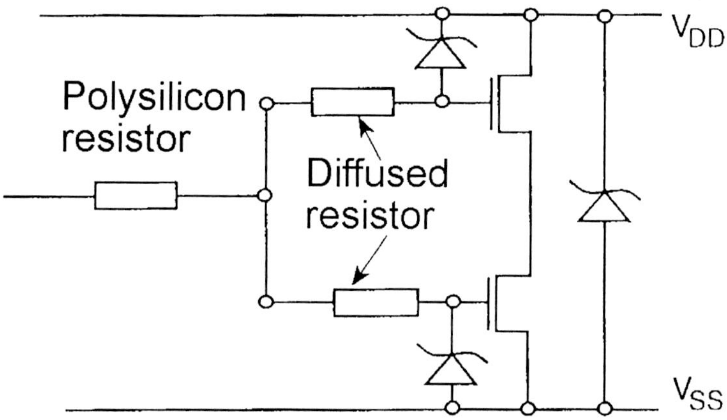

Fig. 138. Schematic illustration of the additional diodes and resistors needed to achieve some protection also against CDM-pulses

Another solution to protect against HBM- and CDM-pulses is given in fig 138.

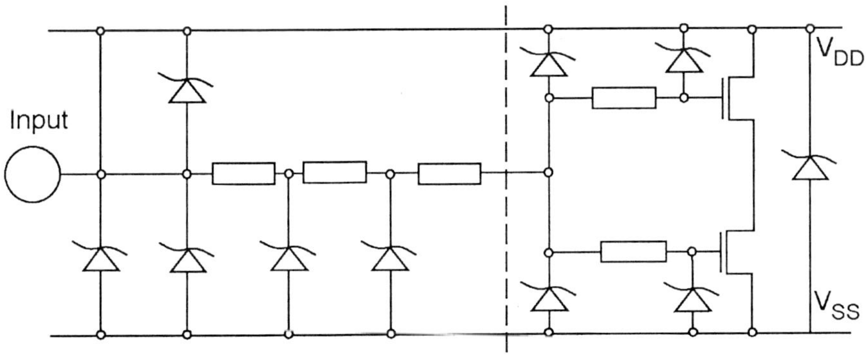

Fig. 139. Input protection circuit, designed to protect against HBM- and CDM-discharges

There are other, rather simple solutions of protection circuits, e.g. a serial resistor (in microcircuits diffused resistors are preferred) followed by a zener diode that decouples the ESD-pulse to earth or to the power supply.

The given descriptions are chosen to give an overview how circuit elements, active and passive, can be combined to achieve sufficient protections against transisents. In the literature (Proc. of ESD and Reliability physics symposia, for instance) many sophisticated solutions of protection circuitry for special cases are found.

Finally, fig 140, some of the first circuits used for ESD-protection are shown. They represent the basic principles for protection and were presented in DOD-HDBK-263 (DOD=Department of Defence, USA).

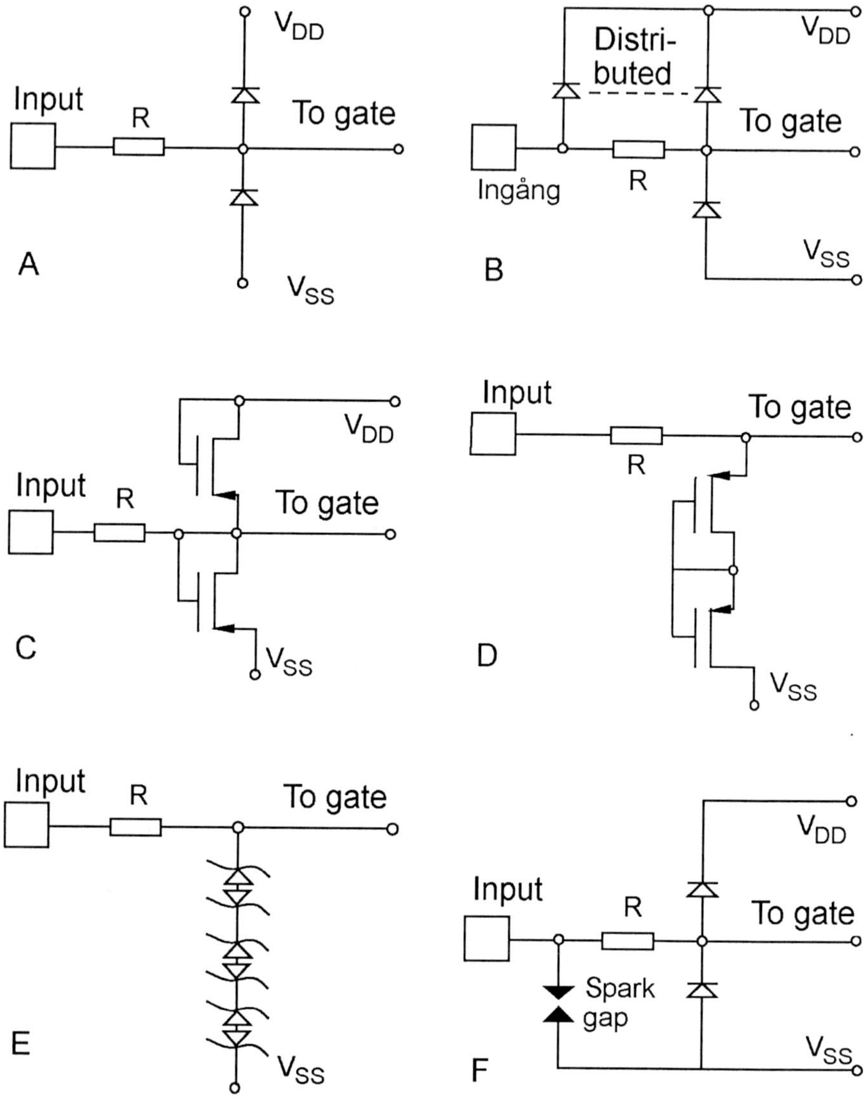

Fig. 140. In DOD-HDBK-263 described simple protection circuits, often used.

A. Diodes	B. Distributed diodes
C. Transistors	D. Transistors, bilateral
E. Zener diodes	F. Spark gap and diodes

In recent years many new protection nets were, as already hinted worked out. They depend very much on the technology chosen for the main circuit, as they are incorporated with it. Scaling down is also a problem. Submicron have become rather common. There can be problems with the energy dissipation of a transient pulse. Optimization is necessary in the design of the protection circuit with regard to other requirements, latch-up, immunity, delay etc. The reader is referred to the literature on this subject, as mentioned before. ESD-sensors to record ESD-events were developed. They can, for instance, be connected to the input of a PCB. Then the ESD-events are monitored.

11.3 Design rules against the impact of ESD

All parts and items shall get proper ESD-protection. At the layout, placement of very sensitive inputs and outputs shall be arranged to avoid contact with parts or elements able to transfer ESD-pulses from the surroundings.

When elaborating the protection circuit it must be taken into consideration that discharges according to different discharge models follow different paths through the component. CDM is the model that correspond to most discharge processes inside a component. In general, components are more sensitive to CDM- and MM-transients than HBM-transients. Because of the increasing use of automization in component mounting the machine model (MM), simulating for instance a discharge from a component tester to the component, is often applied.

If a circuit is scaled down it gets more vulnerable to ESD. This depends on thinner oxide layers, more narrow conductors and more shallow doping wells. The problems can to some extent be circumvented by additional steps in the processing of the circuit, see figure. If in the future some circuits can be protected only to a certain level, say ± 300 V to ± 500 V, an ESD-safe environment must be established to prevent charging to levels as those mentioned. As pointed out before optoelectronic elements can be very ESD-sensitive.

With increased integration the possibility to achieve suitable protection is reduced. Maybe levels of the order of kV cannot be obtained without changing the properties of the circuit, the speed for instance. Some results obtained from testing LSI- and VLSI-circuits will be given. They can be used to arrive at best possible result for the circuit performance.

11.3.1 The thermal problem

A factor of merit for protection circuits was derived together with the protective design in fig 135, by E. Fujishin et al. 1984:

$$F = \frac{E_{\max}}{\tau}$$

E_{max} = maximum energy that can be dissipated

τ = delay caused by the protection circuit

The thermal conductivity inside the component should be as efficient as possible to prevent both melting of metals and oxides and electromigration.

If resistors are used in the protection circuit, they shall preferably be of the diffused type. Polysilicon resistors have no good thermal properties, they can both melt and evaporate. In those cases where polysilicon is used, e.g. out of process technical reasons, the resistor dimensions must be large enough to dissipate developed heat. Resistors are, however, not used frequently in protection for fast circuits as they introduce a signal delay in the circuitry. Not desired current concentrations are avoided by suitable performance of conductors and contacts.

11.3.2
The migration problem

The interface between the contact and the diffusion region is critical, as so called spikes of metal (aluminium) can penetrate into the diffused region and destroy the barrier properties. The diffusion depth is crucial: increased integration results in more shallow diffusion wells. This can be remedied by arranging an n-doped region (fig. 141) under the excessively doped n^+-well.

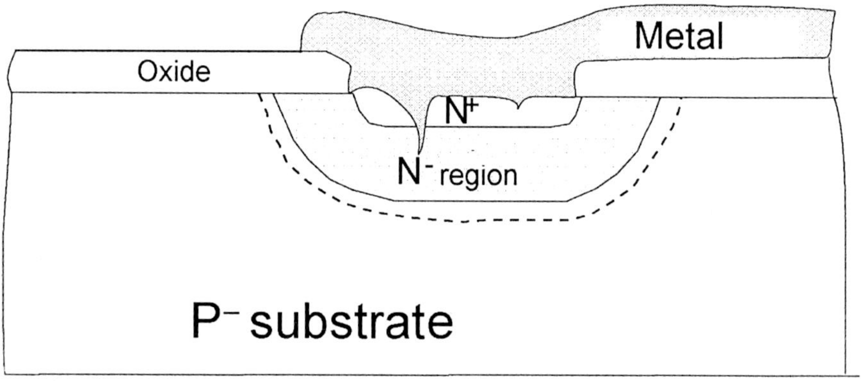

Fig. 141. By arranging an additional n-doped region under the n^+-doped region, a larger marginal against "spikes" without changing the function essentially is obtained

11.3.3
The dimensional problem

It is necessary to emphasize a consequence of increased integration. Beside what was already said about more narrow conductors it should be pointed out that the insulation *between* them is too small to realise a sufficient protection. This implies for the environment that charges must not give rise to voltage beyond ± 300 V!

11.4 Antistatic materials

11.4.1 Background

The phenomenon electrostatic charging and discharge, that was already described, has during the last 20-25 years been paid great attention in electronics. The measures taken to eliminate the influence of discharges, i.e. development of materials that counteract formation of charges and reduce harming effects, has resulted in the growth of a new more or less characteristic area of knowledge, almost a science! Another aspect was application of methods to test components on their vulnerability to ESD.

As already stated all components, even passive, are now regarded as sensitive to ESD. The "new" science has brought a new therminology, the most common terms of which will be given. But first something will be said about different types of materials and their properties.

11.4.2 Specific properties of protection materials

The properties of antistatic materials depend evidently on their electrical conductivity and disinclination to get triboelectrically charged (Chapter 1). The materials shall prevent formation of charges and eliminate those appearing in a gentle way. One more function can be to shield the components against harmful electrical fields. Materials for this purpose has generally a certain volume conductivity or are, provided with a thin metal deposition.

The antistatic materials are commonly characterized by their sheet conductivity. But often the inverse of this value, the sheet resistivity (dimension ohm per square) is applied. Their efficiency is verified by their ability to lead off charges (dissipate charges). For the smooth discharge process of dissipative materials the expression "bleeding off" is used.

Next figure shows, as a first demonstration, how different materials in an electrical field will change it.

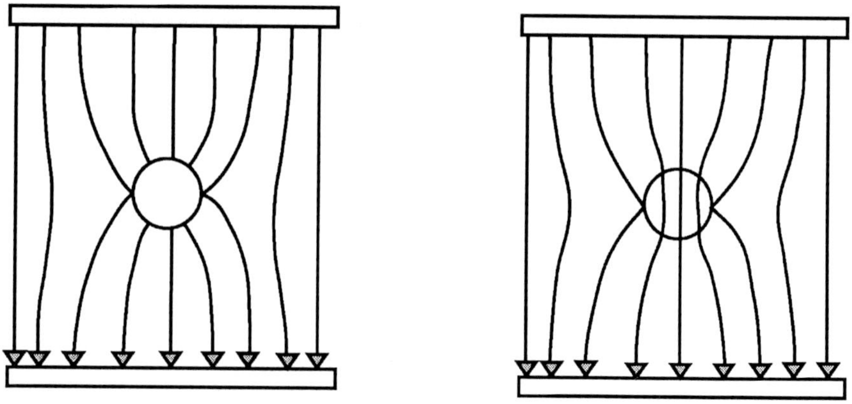

Fig. 142. (A) conductor and (B) insulator (dielectrics) in a similar electrical field. The electric field lines stop at the surface of the metal sphere (A), but penetrate the insulating sphere (B).

The antistatic materials are homogenous, possibly with a conducting surface deposit, or heterogenous, so called multi-layer materials, including a shielding layer as mentioned before (fig 143). The term dissipative (see scales in chapter 1) will be used throughout the presentation. Semiconducting in the sense this concept has in solid state physics is avoided, only for table top (bench) and floor materials it might be applied. The concept antistatic materials is now a common name for all kind of materials counteracting electrostatic discharges and shielding against electric fields. As seen in figure 22 the concept was earlier used for materials in a certain surface conductivity range. Antistatic materials consist often of a basic material, usually plastics as for instance polyethene, with admixtures of conducting substances (e.g. carbon in form of soot).

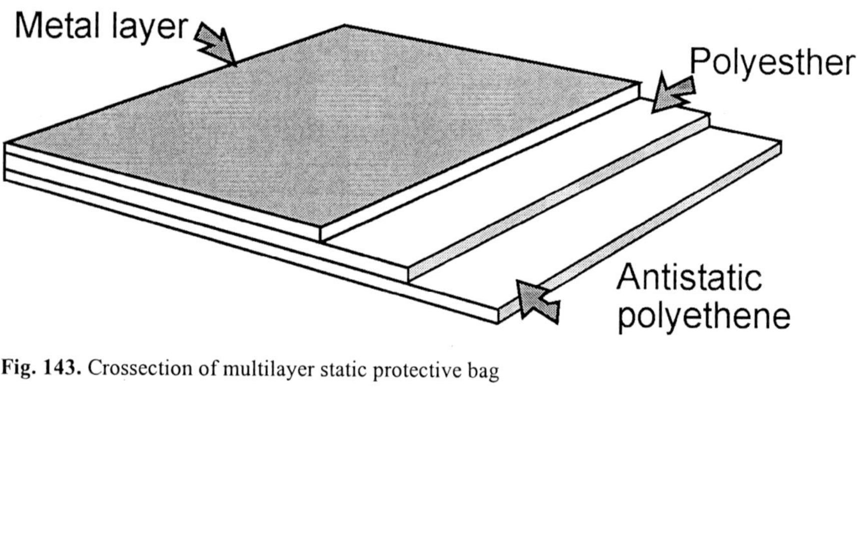

Fig. 143. Crossection of multilayer static protective bag

These materials have varying physico-chemical function and to get a comprehensive survey of them they are divided into three main groups:

A.	Surface deposited antistatic materials
B.	Migrating antistatic materials (see index for migration)
C.	Materials with admixtures to increase the volume conductivity

The research in the field to improve the properties has resulted in many variants, described in literature.

The advantages and drawbacks of the main groups are given together with the description of them.

Multi-layer materials are composed by outermost a conducting metal layer, e.g. nickel, aluminium or chrome, in the middle a monconducting, mechanically strong material and at the inside a dissipating antistatic material. The metal layer can be protected by a thin plastic layer.

As described before, the rise of humidity (RH-value) and ionization of the air are other means to improve the ESD-protection.

11.4.3 Antistatic materials with surface deposition

Substances used: Watersolvable salts, quartery ammonia salts, primary and tertiary amines, phosphoric acid derivates and polyesthers and p-sulphonates.

Manufacturing process: Spraying the basic material with a water or alcoholic solution of the antistatic substance, that also can have admixtures to improve adherence and durability. Alternatively the basic material is dipped (in the form of component tube, for instance) into the solution.

Advantages: The method is cheap and simple, earlier often used to make component tubes (sticks) antistatic.

Disadvantages: The deposition can be dissolved, evaporated or loosened by mechanical friction. The life-length for component tubes of this kind will therefore be only a few months. The antistatic deposition can also, if in contact with other materials, migrate into it and by that cause contamination or corrosion, the antistatic properties getting weaker or cease.

Application: Foil or fibre production, packaging in form of boxes or tubes.

11.4.4 Migrating antistatic materials

Substances used: Components with OH- or NH-groups ionic compounds, amines, amides, esthers, ethers and quartery ammonia salts.

Manufacturing process: Mixed together with basic material before machining process.

Advantages: Improved permanence as the antistatic material migrates to the surface and maintains the antistatic properties. The process is cheap and dyeing is possible.

Disadvantages: The migration can proceed to ambient material and thus cause contamination of the environment.

Application: Counteraction of dust deposits, used for different packaging material as bags and component tubes. The most wellknown of these materials is "pink-poly".

11.4.5 Materials with added ingredients improving the volume conductivity

Substances used: Charbon in form of soot, graphite or fibres, metal powder (aluminium), metal fibres and metalized glass fibres.

Manufacturing process: Mixed together with the basic material before the machining process.

Advantages: Permanent antistatic properties that are independent of humidity, reduced EMI-action.

Disadvantages: Black surfaces, whose durability can be bad. Sometimes expensive to produce.

Application: Antistatic boxes, component tubes, bags, but also protective garments, chair covers and flooring materials. Most wellknown material of this group is "carbon black".

11.4.6 Other antistatic materials

Antistatic materials can also be obtained by electron or UV irradiation of some basic material. This rearranges the molecule and bond configuration to render the material somewhat conductive. The method is, however, rather expensive. The same is valid for the so called electroactive plastics, that are more or less conductive due to their original, interior bond- or electron structure. A drawback is a tendency of instability. Poly-acethylene, for instance, is thermally instable.

During the last two decades many new antistatic materials appeared, new vendors moved into the market. The composition of them can deviate in one or other aspect from the descriptions above. Some of them can be patented.

Pink-poly has been criticized as to its properties and become subject to a thorough evaluation. This material has no shielding effect as "black carbon" has. The reader is referred to the literature for more detailed information on this and other antistatic products.

Untreated expanded polystyrene (EPS), expanded polyethylene (EPE) and expanded polypropylene (EPP) must not be used as packaging material for ESD-sensitive items.

11.4.7
Relative humidity

In chapter 2 it was shown that an increase of the air humidity will counteract charge formation. This is demonstrated in figure 144 with a typical characteristics of the relative charge voltage as a function of the RH-value for some materials.

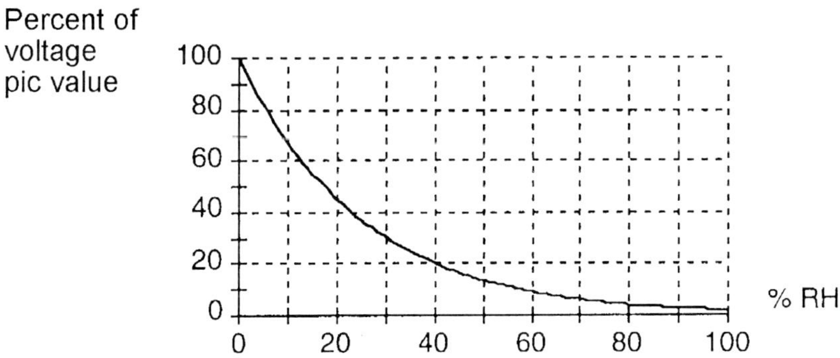

Fig. 144. Mean voltage values of charges in percent of voltage pic value on different materials as a function of the relative humidity (% RH) at room temperature. (Measurements performed at the High Voltage Laboratory, the University of Uppsala, Sweden.)

Followingly it should be preferable to increase the RH-value to improve protection against charge formation. This is also possible for RH-values up to about 50%. In the RH-range 40 to 50% a drastic increase occurs for the so called corrosion index, a measure of the growth of rust on iron. Thus there is a risk of corrosion on equipments for RH > 55%. In areas of very dry climate air humidification is an effective method to limit charge formation. But it will be expensive if big volumes are to be treated.

11.5
Evaluation methods for antistatic materials

The presentation will illustrate basic principles for measurements of antistatic materials. There are detailed international standards and norms for the evaluation of antistatic materials and measurements. References to this are given.

11.5.1
Measurement of flooring and table top (bench) materials

An example of measuring floor and bench materials is shown in figure 145.

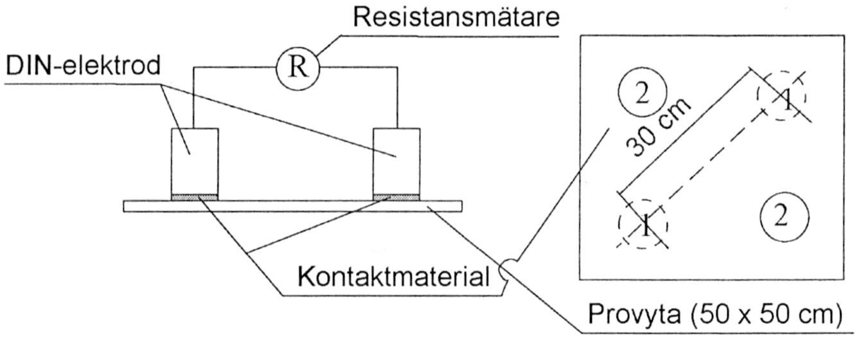

Fig. 145. Surface measurement with two measuring points. Electrode according to DIN-51593

Next figure 146 gives a measurement arrangement to determine the resistance from top to bottom of antistatic mats.

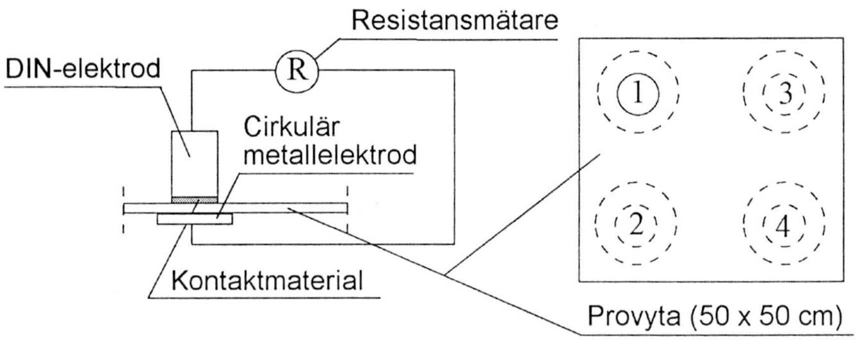

Fig. 146. Top to bottom measurements of mats. Environmental factors in the two cases: $RH = 15^{+0}_{-3}\%$; T = 23 ±3°C. The test surface has four measuring points. Electrode according to DIN-51593

11.5.2 Measurement of triboelectric charging of antistatic materials

In next figure a method to evaluate antistatic bags is given.

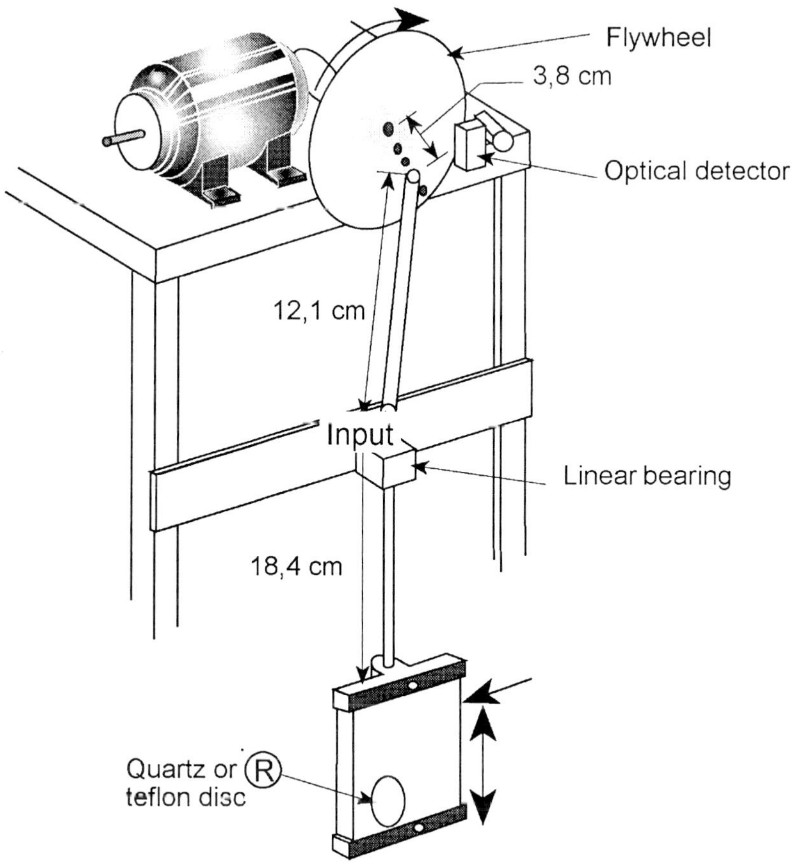

Fig. 147. Arrangement to evaluate ESD-protecting bags according to "Bellcore"-method

The method above was developed by Bellcore (= Bell Communication Research). The disc inside the bag is rubbed against the inner side at the movement up and down. The amount of triboelectrically formed charge is measured in a Faraday pail (fig 147).

Another way to measure triboelectric charge implies the rolling of cylinders made from teflon, quartz, brass etc. downward a surface making an angle with the horizontal plane and covered with the foil to be investigated (fig 148). The charged is measured again in a Faraday pail.

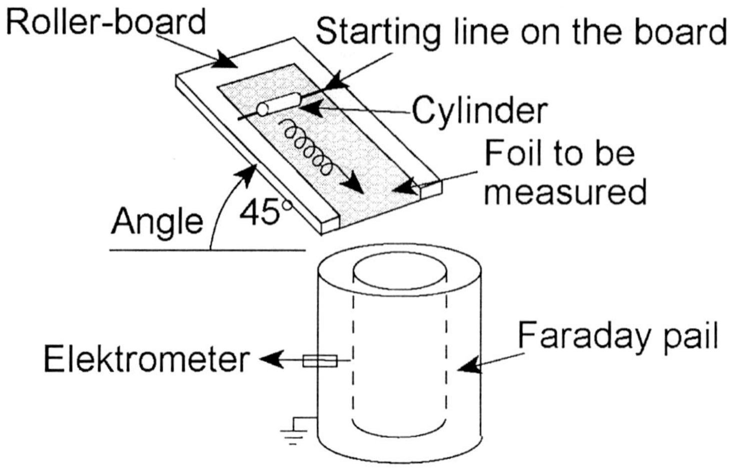

Fig. 148. Arrangement for measurement of antistatic properties of different materials according to a method first applied by the 3M-company

A third method is given in figure 149.

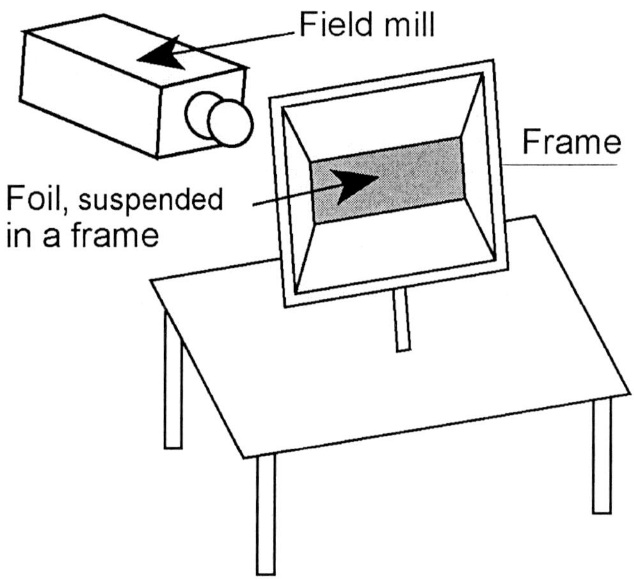

Fig. 149. A foil, charged by rubbing it with a piece of cloth, is suspended in a frame. The voltage of the foil is measured by a "field mill" (chapter 2). Other comparable methods exist for measuring triboelectricity.

The test of materials for component tubes (transport tubes) is described in figure 150.

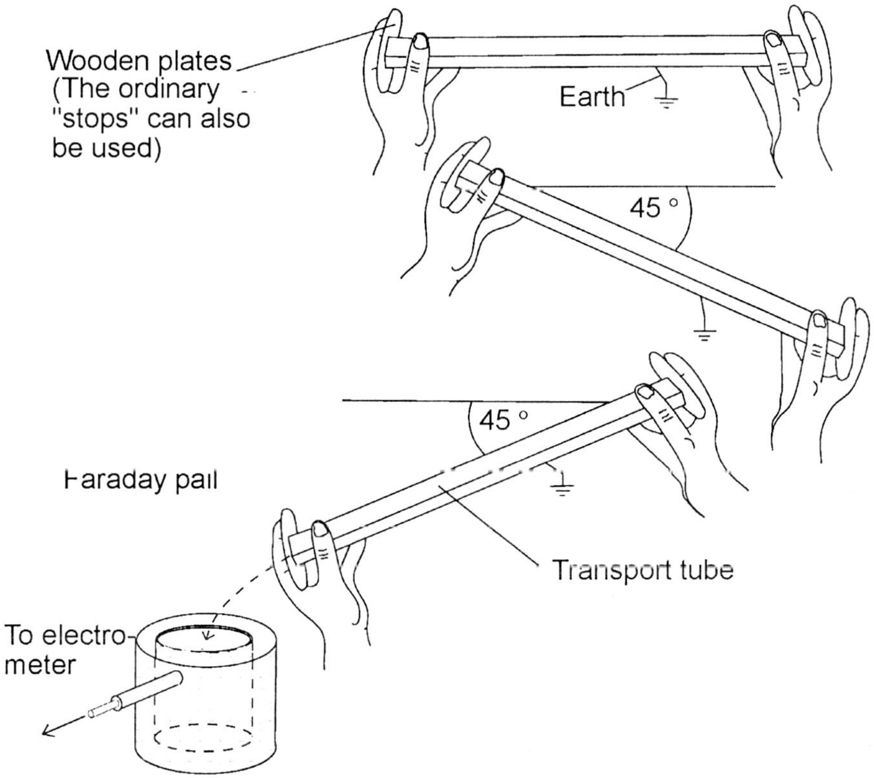

Fig. 150. Testing of transport tubes for IC-components. By tipping the tube with one component inside six times as shown and after the last one, that shall be downward, letting the component slip into a Faraday pail the created charge can be measured

A common requirement for IC-transport tubes is that the created charge shall correspond to not more than 0.03 nC per component pin. Some time the charge requirement is related to the surface size of the encapsulation.

11.5.3 Measurement of surface resistivity

There are many commercially available instruments for measuring the surface resistivity of antistatic materials. A laboratory method can demonstrate the principle (fig 151). Instruments used: A current supply, voltmeter and ammeter together with suitable electrodes. A direct recording instrument is of course to be preferred.

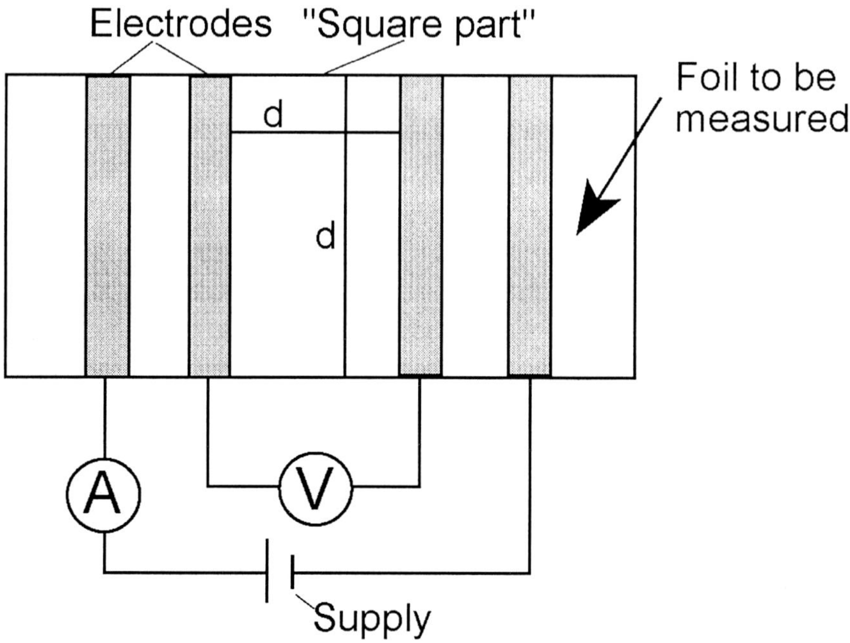

Fig. 151. Arrangement for measurement of surface resistivity. A standard method is given in the US ESD-Ass. standard EOS/ESD-S 11.11-1993 with annular electrodes. Comparable methods are ASTM 257 D, ISO 2878 and IEC 68

Other accessories of antistatic materials as bins, drawers, PCB-cassettes etc., have also to fulfill certain surface conductivity requirements (see EN-100015). Bins and boxes can be tested according to fig 152.

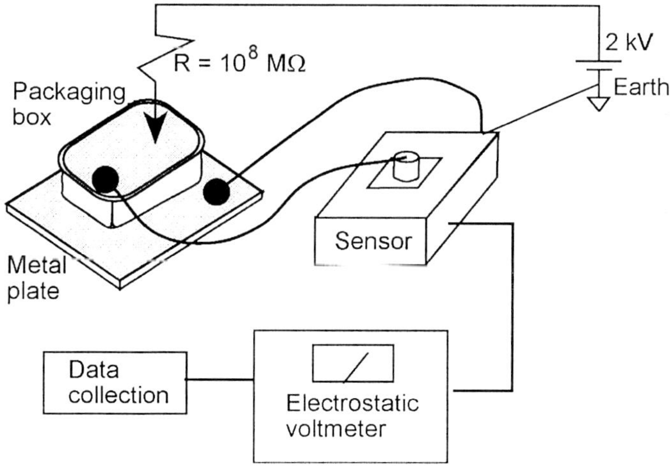

Fig. 152. Testing arrangement to measure discharge from boxes, etc.

Figure 153 shows reults obtained by the described method.

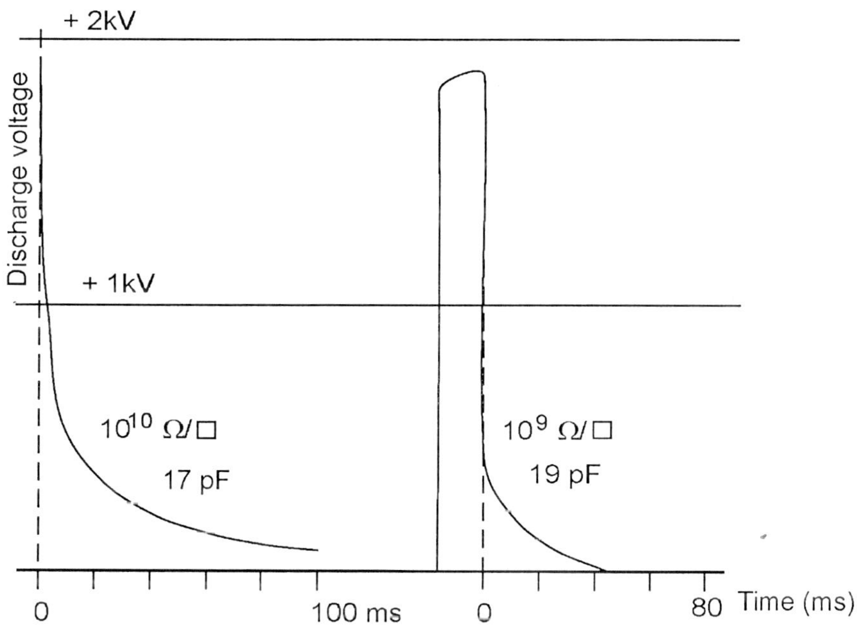

Fig. 153. Rc3ult from diccharge measurement of two boxes with different capacitance and surface resistivity (Baumgartner 1984)

Another way to estimate the static protection properties of materials is to determine the ability of the material to dissipate charges. According to EN 100015-1, antistatic material charged to 10 kV shell discharge to 10 V within 2 seconds.

Commersially automatic meters using a corona discharge to charge a foil to be tested are available. The voltage is measured by a field mill. A measurement is performed from preset starting and end values of the voltage and recording the time between start and end. (Chubb 1992). This method is not recommended for metal foils. Other principles are also used.

11.5.4
Measurement of shielding properties

The methods for measuring triboelectric and shielding properties were the subject of improvements for a long time. To account for the basic principle of testing shielding materials (bags) the method given in EIA-541 is described. This coincides principally with the final standard version for testing shielding materials, US ESD-Ass. standard, ESD S 11.31-1994.

The measuring parts and configuration are shown in figure 154 and 155.

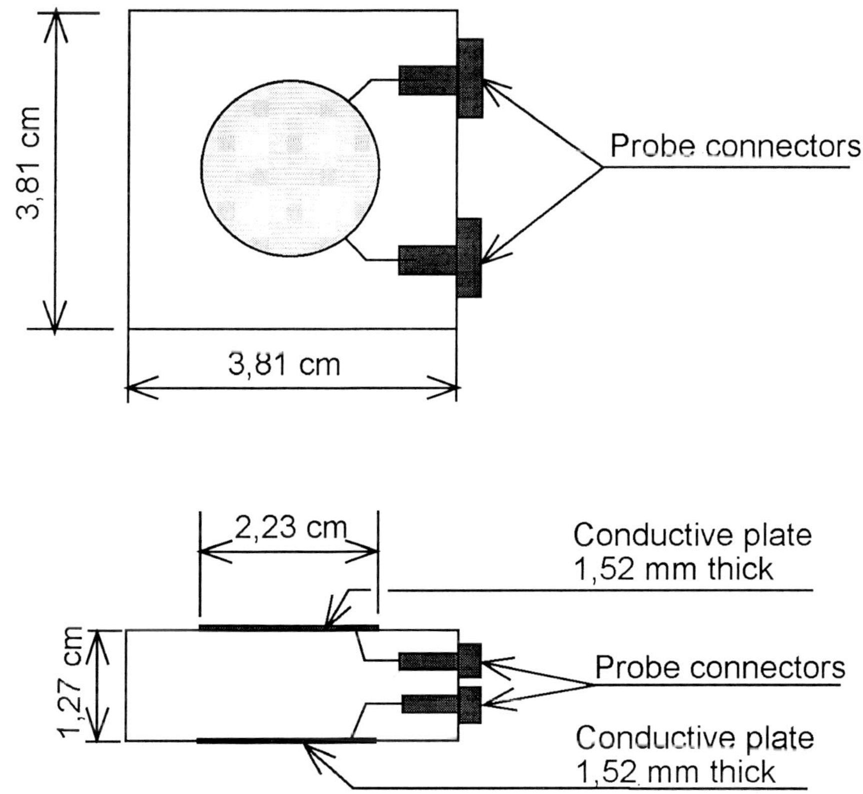

Fig. 154. Capacitive sensor in equipment for measurement of electrostatic shielding properties

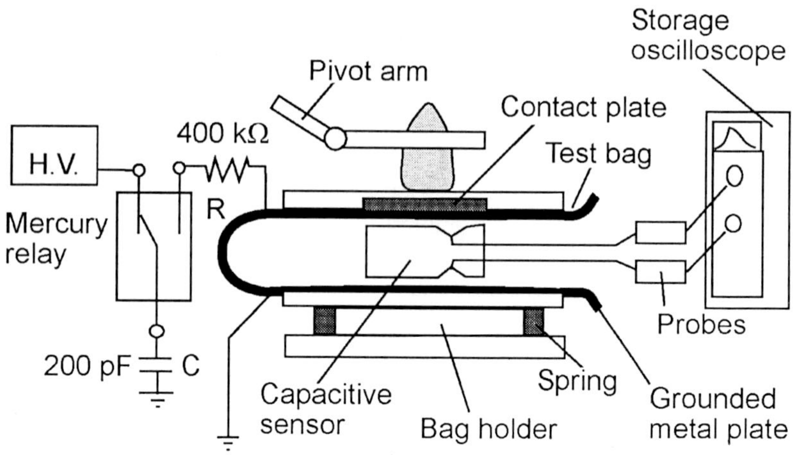

Fig. 155. Test equipment for measurement of electronic shielding properties (EIA - 541)

11.5.5
Certain aspects for choice of antistatic materials

For health and safety reasons following requirements of antistatic materials are stated:

- Not cause corrosion
- Be non-poisonous
- Be inflammable
- Must prevent bacteria and mould growth
- Not cause irritation on skin, lungs or other mucous membranes

There is a great variety of antistatic materials on the market for storing of components and PCB:s. Among the unsophisticated homogenous species there was in the beginning some competition between carbon loaded materials and pink-poly type materials. The former is more conductive (15-20% carbon load) and has some shielding effect, but can cause a short circuit. The migrating liquid in the latter can give rise to corrosion. But they have the advantage to be transparent and not easily form a short circuit. Most antistatic materials are well described in literature. The price will often decide the choice of material. There are special institutes in most countries for certification of antistatic materials.

11.6
The localisation of charges

There are detectors for the localisation of static charges they are in general sensitive to electric fields. But the calibration is given in voltage and polarity. Following figure 156 shows a typical commercial charge detector.

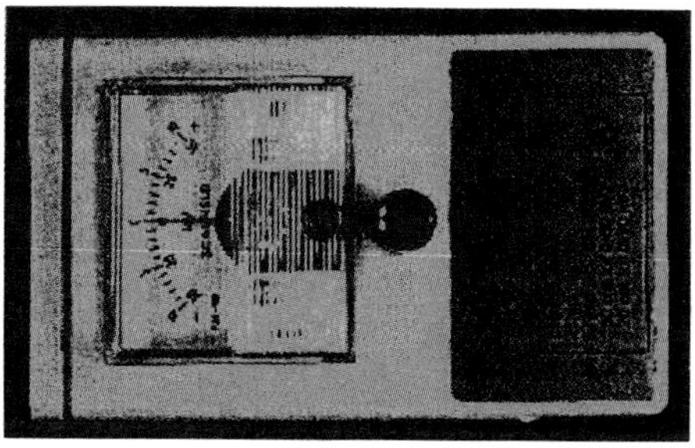

Fig. 156. A commercial detector for static charge localisation

A common principle of instruments used as static charge detectors is demonstrated by the circuitry in figure 15. Method to fix zero point is included.

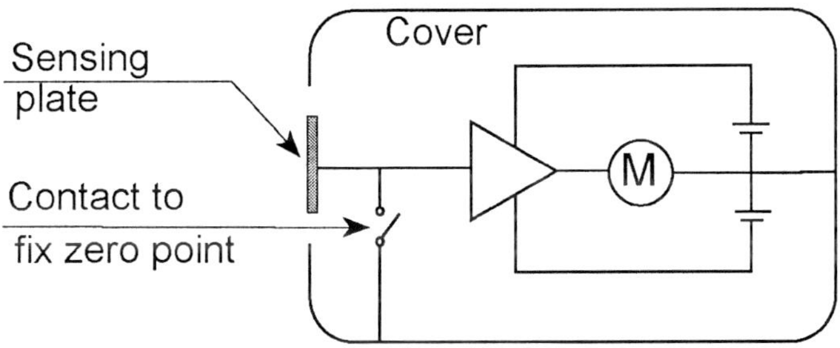

Fig. 157. Working scheme of static charge detector. An electric field from charges in the environment induces charges on the sensing plate. The DC-amplifier transforms the charges into an electrical signal registrated by the measuring instrument M. The instrument gives also the polarity. A way of calibration of the detectors is described in figure 158.

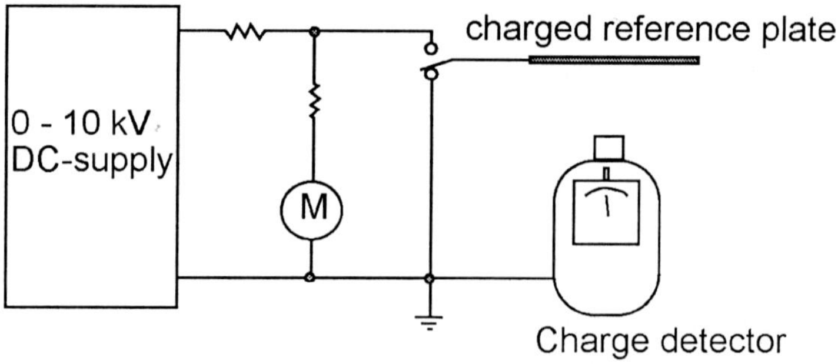

Fig. 158. Calibration arrangement for static charge detectors. The calibration is dependent on the size of the reference plate relative the size of the sensing surface and the distance between them. In this way, as the calibration arrangement used by the manufacture can be different from the measuring conditions in application, the accuracy is lowered

There are instruments of good precision and well-defined probes, that make measurement with high accuracy possible.

A source of error when using charge detectors is pointed out in figure 159.

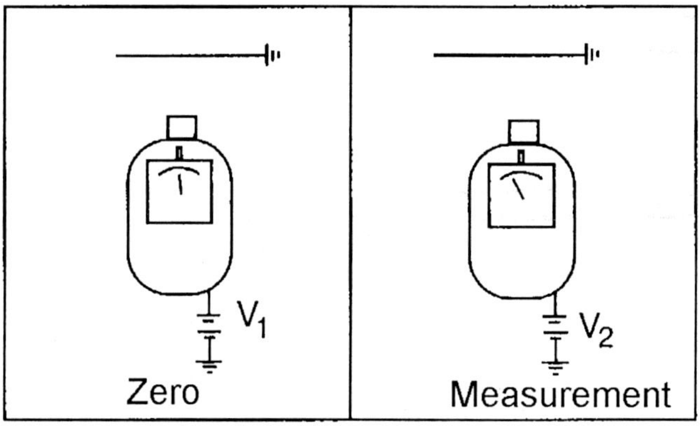

Fig. 159. Error in static detector application originating from different reference voltages, i.e. one reference at calibration procedure another in application (Lonborg 1983)

11.7 Protection program

From the preceding presentation it is quite clear that the components as far as possible must be protected against ESD caused by the operators and surrounding objects, also inductive fields.

The experience of ESD in electronic industry gave rise to introduction of protection program, including teaching about ESD problems. Many programs were described in literature, dealing with handling rules and suitable antistatic materials. The "awareness" problem was early realised. People must be aware of formation of electric charge and that they are a hazard in electronics. If somebody feels or hear a discharge he understand the risks. Discharges less 3 kV are not felt, but can harm electronics. But to demonstrate with discharges on components and show the damages of them is convincing.

At handling sensitive components it is very important that the operator is earthed by a wriststrap whenever it is possible. Working benches shall have dissipative table top cover (table mat) earthed by a protective resistor. Alternatively, or as redundancy, dissipative floor mats the personnel wearing dissipative shoes. Chairs shall be covered with dissipative material electrically in connection with the floor. The same is valid for trolleys. Tools shall have antistatic handles and kept in contact with the earthed table mat when not used.

According to the EN 100015-1 norm (see p. 167) shall antistatic smocks where applicable be used. It must be observed that cotton should not be used in clean rooms as cotton forms naps.

At a rather early stage in the planning of protection program following twelve key factors for a sucessful program was stated (Dangelmayer).

- An effective plan of implementation is absolutely necessary
- Commitment of the management is important
- The activities requires a full-time coordinator
- An active ESD committee is to be selected
- Realistic ESD-requirements must be formulated: Adaptation to the susceptibility of components the three ESD-models HBM, CDM and FIM must be taken into account
- Awareness - training for measurable goals
- A critical evaluation of the activity
- Establishment of suitable resources for ESD-testing, etc
- Make the communication with involved people effective
- A systematic planning of the activity
- Take human behavior into consideration
- Continuously introduce improvements of the ESD-protection

Also when working on system level ESD-protection must be used. At installation, testing, control and maintenance, there shall be dissipative floors (floor mat), dissipative shoes or heelstrap and/or wriststrap.

By PCB-exchange and tests wriststraps are to be used. Installation or change of magazines can be executed without wriststrap only there is a dissipative floor mat and dissipative shoes/heelstraps are worn. If the wearing of a wriststrap is inconvenient to the operators and the flooring is not dissipative, a dissipative earthed mat is to be used. Trolleys must have conductive wheels!

To achieve a sufficient protection for microcircuits against ESD they must also be supplied with protection circuits incorporated with the main circuits. A complete ESD-protection is shown in figure 160.

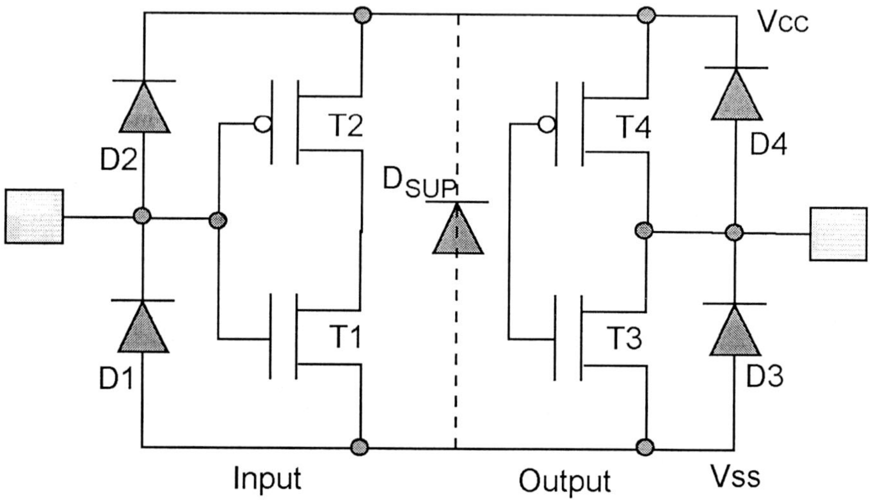

Fig. 160. Complete ESD-protection for a microcircuitry. Both in- and outputs are protected with special protection circuits

11.7.2
An ESD-protected workstation (EPA)

Figure 161 shows an ESD-protected working place, called EPA in EN 100015-1

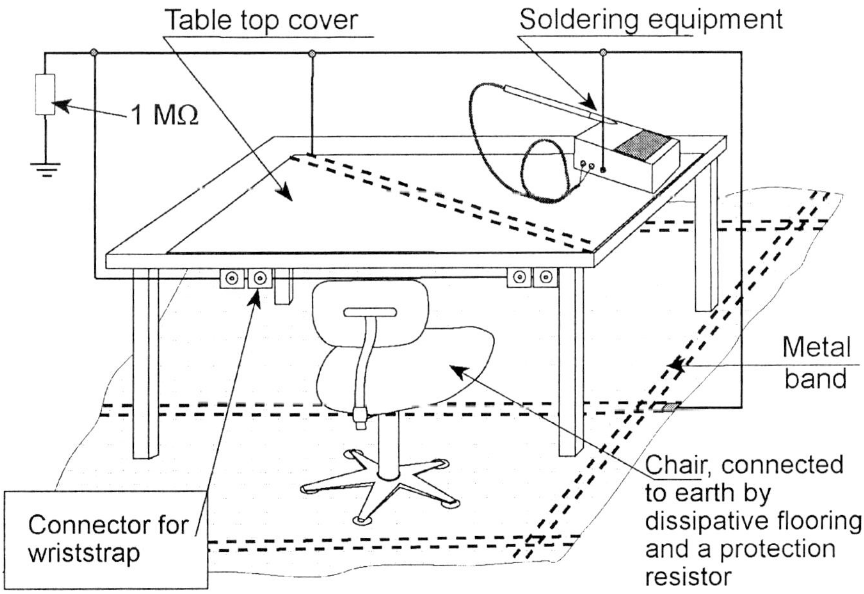

Fig. 161. The figure shows an example of how an ESD-protected working place can be performed

Wriststraps are very important, a description of their main parts is given in next figure 162.

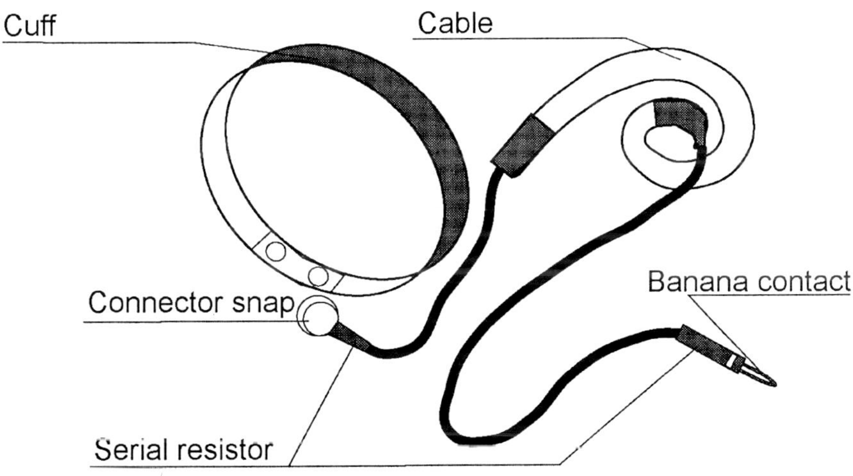

Fig. 162. Example of a wriststrap

11.7.3 Ionizers

When handling very ESD-sensitive components, optoelectric circuits for instance, an ion generator (ionizer) can be used. In clean rooms so called whole-room ionization is sometimes applied. There are different processes for creation of ions. Figure 163 demonstrates "field ionization". When the electric field becomes strong enough breakdown starts in the air with formation of ions and lighteffect, a "corona" is created.

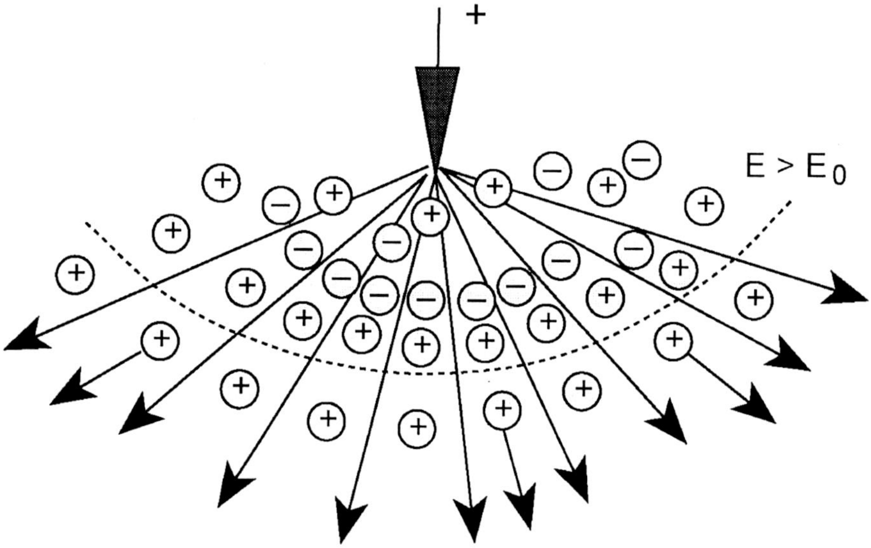

Fig. 163. Electric field ionization ("corona" starts at $E>E_0$)

Other performances of ionizes are given in figure 164.

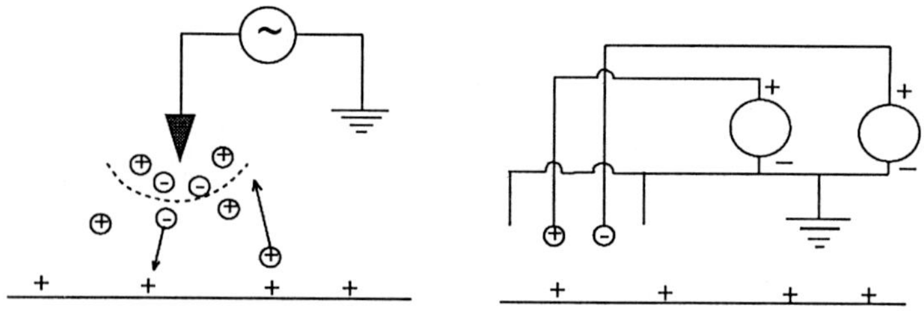

Fig. 164. Charged material

AC-ionization Double DC-ionization

The ions neutralize created charges on surrounding surfaces. For bipolar ionizer there must be a balance between pos. and neg. ions. There is a risk of corrosion of the electrodes with contamination as consequence. The system must be served.

The ion current can be measured as shown in figure 165.

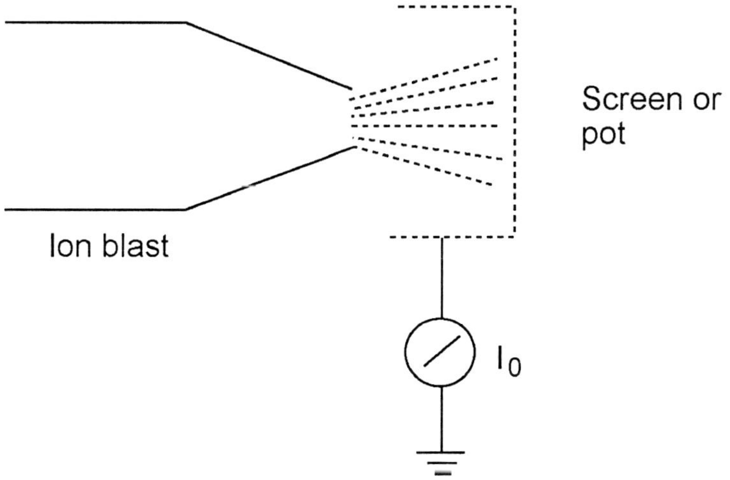

Fig. 165. Measurement of the ion current

Ion generators using radioactive particle radiation (e.g. α-particles), are in general not approved to use, though they are very effective. Accidents with them resulted in their replacement with corona system ionizer (unipolar or bipolar). For AC-ionizer the frequence can have an influence.

Low frequencies (~Hz) can prevent that the positive and negative ions recombine before reaching the charged surface. For the same reason a "pulsed" ion source is sometimes used. Next figure 166 shows a commercial ionizer.

Fig. 166. A portable ionizer that can cover a limited surface. Sometimes the ionizer is supplied with a fan to blow out the ions. The fan air is warm for personal comfort

11.7.4
Example of a whole room air ionization system

Figure 167 gives the arrangement of the ionization system. A foil for measurement is first charged by means of an ion-beam generator to about 5 kV. At that moment the ionization system is switched on. Then time elapsed until the voltage of the foil has dropped to 50V is recorded. Normally a discharge time of 10s is accepted.

The ionization level is regulated by means of a chargable metal plate visible to the left on the bench in figure 167. The electric potential increase due to the presence of ions is measured on the bench, figure 167 and a maximum, allround value of it is chosen, e.g. 20V, by adjusting the corona formation in the ionizers. If a balance between positive and negative ions is not obtained surrounding material will get charged! The same metal plate as before can be used. It is first charged and the potential remaining after start of the ionization system is measured. The magnitude of the rest potential is an indicator of the balance obtained (zero means exact balance). Earthed surfaces near the ion source shall be avoided as they attract the ion beam.

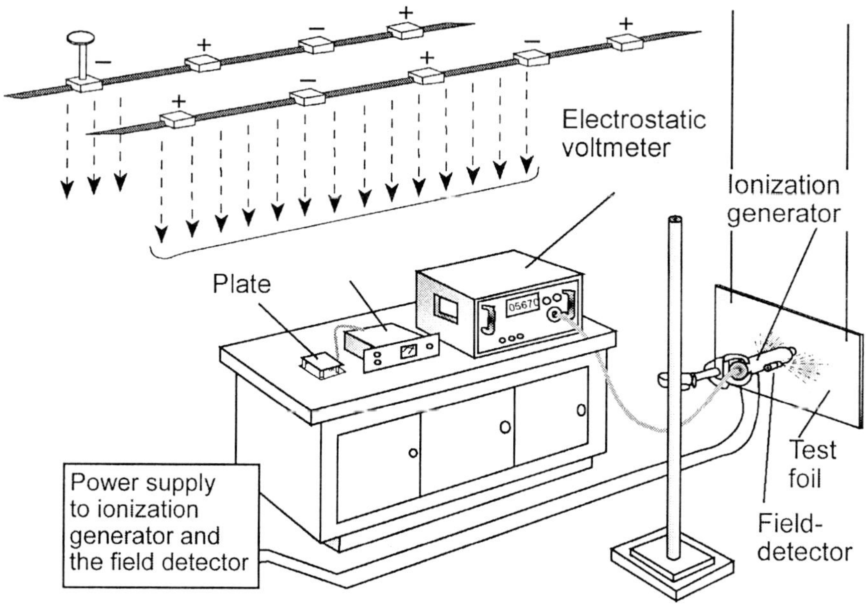

Fig. 167. Arranagement to determine the efficiency to a whole room air ionization system. The ionizer generate alternatively positive and negative ions. (Mykkanen and Blinde, 1983)

Instead of "wires" as in the figure 167 commercially available rods with corona generators can be used, placed in the air-boxes of the ventilation system. Several types of ionizers are available commercially at different prices. Many applications of ionization are described in the literature.

11.8 Norms and standard

11.8.1 CECC-, IEC-, ISO- and EN- norms

CECC and EN are European norms, while IEC and ISO stand for global norms (valid for member countries).

IEC 1000-4-2 is a norm for IT-equipment (e.g. telecommunication).

The following European norms are valid for the ESD-protection areas. A parallell norm work is going on within IEC.

EN 100015-1 General requirements (subject to update)
EN 100015-2 Requirements for Low Humidity Conditions

EN 100015-3 Requirements for Clean Room A
EN 100015-4 Requirements for high Voltage Environments

A paramount norm is ISO 9000 Quality Control. The US ESD-Ass. standard Committees have elaborated standard for antistatic materials and component testing.

The following are referred to:

EOS/ESD	S 1.0-1987	Wrist straps
EOS/ESD	S 4.1-1990	Work surfaces-Resistive Characterization
EOS/ESD	S 6.1-1991	Grounding-Recommended Practice
EOS/ESD	DS 7.1-1994	Revised-Floor Materials-Resistive characterization of Materials
EOS/ESD	DS 54.1-1996	Floor materials and Footwear-Resistance Measurement in combination with a Person
EOS/ESD	S 11.11-1993	Surface Resistance Measurements of Static Dissipative Planes Materials
ESD	DS 11.31-1993	Packaging-Shielding Bags
EOS/ESD	S 3.1-1991	Ionization
ESD	ADV 3.2-1995	Selection and suceptance of Air Ionizers
ESD	ADV 53.1-1995	ESD Protective Workstations
ESD	S 9.1-1995	Footwear-Resistive Characterization
EOS/ESD	DS 2.1-1995	Garments

For testing of component ESD-sensitivity:

EOS/ESD	S 5.1-1993	Human Body Model (HBM)-Component level
ESD	DS 5.1-1997 (revised)	Electrostatic Discharge Sensitivity Testing: Human Body Model
ESD	S 5.2-1994	Machine Model (MM)- Component level
EOS/ESD	DS 5.3-1993	Charged Device Model (CDM) Component level

The following publication:
Electrostatic Discharge Susceptibility Data (VZAP), is obtainable from:
Reliability Analysis Center
P.O Box 4700
Rome, N.Y. 13442-4706
USA
It contains a variety, of data from ESD-testing of electronic components

11.9 Labels for marking

According to EN 100015-1 packaging containing ESD-sensitive parts shall be marked with following symbols and provided with a warning text, fig. 168, 169.

Fig. 168. ESDS, basic symbol ESDS, simplified symbol

Areas for handling of ESD-sensitive items are recommended to be provided with a warming text

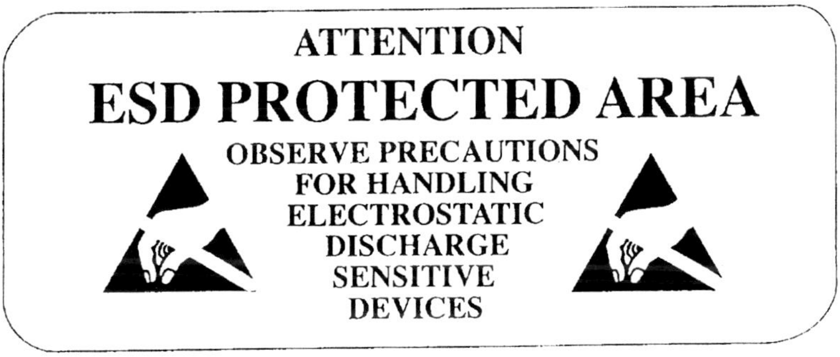

Fig. 169. Warning mark for EPA. The label must have a minimum size of 300x150 mm

It is prescribed that within an EPA separated ESD-earth (EEF), earth bonding point (EBP) and earth grounding point (EGP) shall be marked:

Fig. 170. Examples for marking EEF. Corresponding texts for EBP and EGP will be "Earth Bonding Point and Earth Grounding Point"

12 Failure frequency and costs for ESD

Information about the frequency of ESD-failures, both in production and field use, is of great importance. Due to high costs for the necessary, time - consuming failure analysis sufficient data can be difficult to get.

Failure hazards can be related to different situations for the components, arrival inspection for instance. This is given in figure 171 together with relative cost for repair.

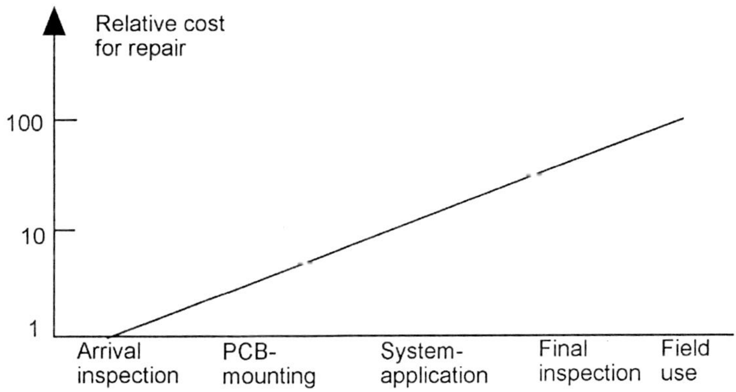

Fig. 171. Estimate of costs for ESD-failures at different stages in the production

12.1 Failure occurrance

In telecommunication the starting up period of a switching system, that takes about one year, is the most sensitive operation time of it. There are many statements about how many of the failures during this period are due to ESD, but 50% is a figure of order, then contribution from latent failures are taken into account.

The normal, mean ESD-failure rate is related to the degree of ESD-protection applied. Even if the failure rate is low, due to effective protection, an improvement can still reduce the costs of ESD-failures in the case of large amounts of components. A computer manufactures that had as low failure for a logic circuit, found it repayable to improve the ESD-protection, though only some 0.01% was gained! The great amount of components made it profitable. The failure rate of electronic components during different periods of their use is usually described by a so called bath-tub curve as in figure 172.

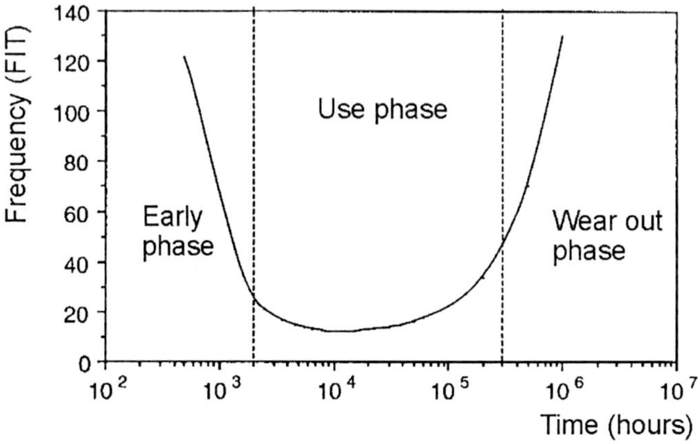

Fig. 172. Failure frequency as a function of time also called "bath-tub curve" for modern active components

The different phases of the curve also indicates what type of activity that gives rise to the failure frequency.

Early failures come from quality dismisses as non-hermetic encapsulation with moisture leakage into it, failing bonds, material damages etc. In this group are also failures due to handling of components as among others ESD-damages.

Failure intensity $Z(t) = \dfrac{f(t)}{1 - F(t)}$

When f(t) = probability density
F(t) = failure distribution function = $\int f(t)dt$.

Early failures disappear normally within 3000-4000 h. The failure intensity is then usually anticipated as constant $z_I=\lambda$. See figure 173.

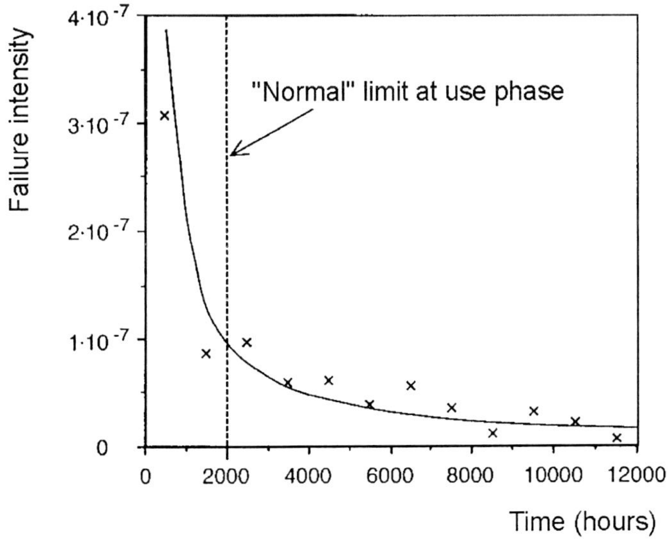

Fig. 173. Failure intensity z(t) as a function of time, recorded from field use of active components in "benign" environment. The failure intensity was defined as $\frac{n}{N}\Delta t$. Where n= number of failures N=number of components and Δt= time interval, here = 1000 h. Assuming a limited number of early failures, the probability density can be approximated with an exponential function $f(t)=\lambda e^{-\lambda t}$, giving the failure intensity $z(t)=\lambda$. The investigation made for 25000 h showed only few other failure beside these early ones. It is reasonable to be wondering if not the handlling of components, that often gives ESD-damages is not the largest source of failures in modern components, especially in "benign" environments!

Often a so called burn-in process is made to reduce the early failures. This mean that the components are operated at elevated temperature and sometimes also elevated humidity for some days. But it is questionable if the number of failure decreases, because a lot of handling of the components are done in the procedure increasing the possibility of ESD-failures! The procedure requires an effective ESD-protection program.

In the *use-phase* the failures are due to the "normal" stress the component is submitted to in its application and the degree of active operation. These failures are random. ESD-failure occuring in this phase are of different origins: latent failures, especially time-dependent, failures appearing at transport, retrimming or other kind of handling. It is difficult to select different types of failure in the use-phase.

The failure intensity of this phase is usually the base for life-time predictions of equipments. The probability density is approximated with a functions: $f(t) = \lambda \cdot e^{-\lambda t}$ giving a failure intensity: $z(t) = \lambda_S$, constant in time.

In the *wear-out* phase the normal ageing process of the components is the cause for failures. This process is due to different types of fatigue as electromigration, successive displacement of threshold values caused by "hot electrons" and phase transformations in materials. ESD-failures do normally not appear. ESD-damages with electromigration arrive much earlier as "early dead components". The expected life-time of components passing the two first phases are "normally distributed" with mean life-time μ and dispersion σ i.e. $N(\mu,\sigma)$. The failure intensity z_{WO} is followingly not constant.

There has been put forward a suggestion, that for comparison between different groups of components a simplified version of the bath-tub curve, based on parameters λ_I and λ_W should be used. λ_I = failure intensity for the initial period and λ_W = failure intensity during the wear-out period. For the use-period $\lambda_S = 0$.

ESD-failures are often of sporadic and random character, but can then result in a high loss of components. As an example a type of micro-processors from two different vendors can be mentioned. In the arrival inspection 5-15% were failing. Analysis showed that a "clock-pin" without ESD-protection became damaged.

ECL-circuits are sometimes sensitive to ESD, causing troubles. In Japan it was observed that ECL-components, probably due to latent failures, cause intermittent failures. The failing components were difficult to localize!

Likewise ECL input circuits of logic circutis have failed due to ESD. But it is difficult to damage these components by HBM-stresses. Later failures were, however, revealed at temperatures check testing. Handling of components are likely to introduce latent failures, that, for instance, by discharge from a charged cable transform them into catastrophic failures! That process corresponds to a discharge according to the machine model (MM). Testing the components in a MM-tester can confirm this.

Investigation of failures appearing in field use are rather rare. Two important institutions working on reliability problem presented results given in table.

Source	US-MIL-HDBK-217F			RDF91			Mean value (%)
Environment	GB	GF	GM	1)	2)	3)	
EOS/ESD	71	62	23	42	45	27	45
Encapsulation	23	26	42	8	33	40	29
Chip	6	6	8	50	22	33	21
Other	0	6	27	0	0	0	6

Table 7.

Field failures in procent from two institutions, RADC (US-MIL-HDBK-217F) and British Telecom (RDF91).

GB = office and laboratory environment, investigation of bipolar circuits,

GF = permanent installations in unwarmed buildings, IC in DIP

GM = installations on movable supports (lorries), 523 bipolar and 123 MOS SSI/MSI.
Interface bipolar SSI
Interface bipolar MSI
Interface CMOS MSI

It is not known what ESD-protection was applied at the manufacture of the PCB:s upon which the components were mounted.
It cannot be interpreted that all failures belonging to EOS/ESD are ESD-failures. Two earlier investigations one in Sweden and one in USA, showed about 75% were ESD failures of the EOS/ESD failure rate. Some of the other 25% failures can also have started as minor ESD-failures!

Another interesting observation is, that the percentage of EOS/ESD-failures decreases when the environment gets more severe this can be interpreted as other failures than EOS/ESD start appear in such environments, while the number of EOS/ESD-failures is rather constant. This agrees with the observation, that most ESD-failures are discovered in connection with trimming or burn-in of equipments. In next figure the trend of failure is given.

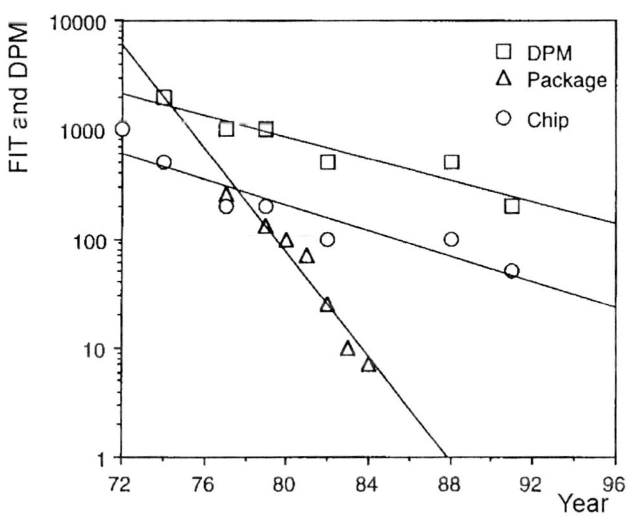

Fig. 174. Trends of failure intensities on chips and encapsulations, unit FIT = failures per 10^9 hours. The appearance of early failures measured in DPM = defects per million items is also given.

An estimate of early failures from figure 174 gives 10^3 DPM for 1980, i.e. 1‰. This shall be compared with a complementing investigation giving 8‰ and done with CMOS-circuits manufactured 1980, see also figure 174.

Corresponding trend for chip failures gives λ_{chip} (1980) = 250 FIT and for encapsulation (1980). λ_{enc} (1980) = 100 FIT what gives for the completing investigation 1.25 respectively 0.5‰ failures. This gives together 2.75‰ failure for the first 5000 hours.

Two conclusions can be made from the complementing investigation:

Failures caused by encapsulation misfits and ESD are evidently under-estimated

Failures due to ESD is probably still the most likely cause to early failures.

By allowing at handling and measurement of a certain component some have a total ESD-protection and other none, the importance of the protection can be estimated. The following table shows a case with a hybrid-circuit (HIC). (Dangelmayer, 1983).

Relation unprotected/ESD-protected	
Defects in process-test	1.9 times
Defects in final test	2.4 times
After handling in unprotected packaging	
Defects after test	5.5 times

In the unprotected the circuit failed 5.5 times more than with ESD-protecting packaging.

To estimate the frequency of ESD-failures among the defected parts a failure analysis was performed with following result:

No failure	48%
ESD-induced failure	39%
Other failures	13%

12.2 Cost of ESD-failures

The following account for implementation of an ESD protection program is rather old (Lockhead 1979-1982) but gives a good idea about the cost relation.

Cost of implementation:

Bench protect	$ 48 000
ESD-protection boxes	$ 21 000
Wriststrap/footstraps	$ 4 000
Semiconducting floor	$ 100 000
Ionizers	$ 2 700
Guideline, documentation	$ 2 000
Training of ESD-watching	$ 2 400
Control procedure	$ 5 000
Sum	$ 185 100

Cost of up-date and maintenance:

Articles of consumption	$ 20 000	
Training courses	$ 4 000	(100 x 2 hours)
ESD-control	$ 5 000	
Remaining ESD-failures	$ 16 000	
Yearly cost	$ 45 000	

Saving by avoiding ESD-failure:

Eliminated ESD-failure	$ 1 800 000
Cost of implementate	$ - 185 000
Saving year 1	$ 1 615 000

Savings during four years by avoiding ESD-failures:

Saving	year 1979	$ 1 615 000
Saving	year 1980	$ 1 775 000
Saving	year 1981	$ 1 790 000
Saving	year 1982	$ 1 800 000

There are estimates that the electronic industry might loose 10 billion/year in USA because of ESD-damages (Source: Stonhard Inc., 1989). This corresponds to 5-10% of the production value.

13 Index

14 References

The references are chosen as being typical for the subject or recently published. Valuable references are found in the Proceedings of the annual EOS/ESD-Symposia in USA, the annual Reliability Physics Symposia in USA, Journal of Electrostatics, the annual European Symposium Reliability of Electron Devices – Failure Physics and Analysis, the IEEE Transactions on Electron Devices and other journals and publications.

14.1 The discovery of static electricity, and its manifestation

Catalogue of Books and Papers in Electricity and Magnetism,
Ekelöf S, Belonging to the Institute for History of Electricity Chalmers University of Technology, Göteborg, Sweden 1991

Static and Dynamic Electricity,
Smythe W R, McGraw Hill, N.Y., 1968

Estimation of Discharge Energy Released from Charged Insulators,
Tabata Y, Proc. 1984 EOS/ESD-Symp.. p. 50, 1984

Atmospheric Electricity – Nature's Own ESD,
Jonassen N, Proc. 1990 EOS/ESD-Symp. p. 1, 1990

Electrostatic Discharges. A charge driven phenomenon,
Greason W D, Proc. 1991 EOS/ESD Symp. p. 1, 1991

Do Gases Charge?
Jonassen N, Proc. 1994 EOS/ESD-Symp. p. 221, 1994

Explosions and Static Electricity,
Jonassen N, Proc. 1995 EOS/ESD-Symp. p. 331, 1995

Electromagnetic Theory,
Hallén E, Chapman and Hall, London, 1962

ESD – A Decade of Progress,
McAteer O J, Proc. 1988 EOS/ESD-Symp. p. 1, 1988

14.2 Relation to Electronics

IEC 721-2-1, Temperature and Humidity,
Mollierdiagram

Static Electricity a Different Kind of Contamination,
Walker R C, Microcontamination, Oct/Nov 1983

Static-Electric Characterization of Semi-Insulating Materials,
Jonassen N, Proc. 1984 EOS/ESD-Symp. p. 45, 1984

Coupling of ESD-Generated EMP to Electronics,
Boverie, B, Proc. 1988 EOS/ESD-Symp. p. 173, 1988

Electrostatic Damage in Electronics: Devices and Systems,
Greason W D, John Wiley and Sons, N.Y., 1987

Electrostatic Decay and Discharge,
Jonassen N, Proc. 1991 EOS/ESD-Symp. p. 31, 1991

14.3 ESD-testing and models

The Characterization and Control of Leading Edge Transients from Human Body Model ESD Simulation,
Chemelli R G and De Chiaro L F, Proc. 1985 EOS/ESD-Symp. p. 155, 1985

Theoretical and Empirical analysis of the Effects of Circuit Parameters on the Calibration of HBM ESD Simulators,
Bilodeau T, Proc. 1989 EOS/ESD-Symp. p. 43, 1989

A programmable Equipment for Electrostatic Discharge,
Shaw R N and Enoch R D, Proc. 1983 EOS/ESD-Symp. p. 48, 1983

An Experimental Investigation of ESD-Induced Damage to IC's on Printed Circuit Boards,
Shaw R N and Enoch R D, Proc. 1985 EOS/ESD-Symp. p. 132, 1985

The Dependence of Electrostatic Destruction Voltage on Device Structures of pn-junctions and Insulated Films,
Maeda M and Wada T, Proc. 1989 EOS/ESD-Symp. p. 157, 1989

Beyond MIL HBM-Testing: How to Evaluate The Real Capability of Protection Structures,
Avery L R, Proc. 1991 EOS/ESD-Symp. p. 120, 1991

Degradation in In GaAsP Semiconductor lasers Resulting from Human Body Model ESD,
De Chiaro L F and Unger B A, Proc. 1991 EOS/ESD-Symp. p. 172, 1991

Standard ESD Testing of Integrated Circuits,
Van Roozendal L et al, Proc. 1990 EOS/ESD-Symp. p. 119, 1990

Analysis of HBM ESD Testers and Specifications Using 4th Order Lumped Element Model,
Verhaege K et al, Proc. 1993 EOS/ESD-Symp. p. 129, 1993

ESD Damage from Triboelectrically charged IC Pins,
Bossard P R et al, Proc. 1980 EOS/ESD-Symp. p. 17, 1980

Charge Device Model Testing: Trying to Duplicate Reality,
Avery L R, Proc. 1987 EOS/ESD-Symp. p. 8, 1987

A field Induced Charged Device Model Simulator,
Renninger R G et al, Proc. 1989 EOS/ESD-Symp. p. 59, 1989

Mechanisms of Charged-Device Electrostatic Discharges,
Renninger R G, Journal of Electrostatics, Vol 28, No 3, 1992

Influence of Tester Parasitics on "Charged Device Model" Failure Thresholds,
Gieser H and Egger P, Proc. 1994 EOS/ESD-Symp. p. 69, 1994

Influence of Tester, Test Method and Device Type on CDM ESD Testing,
Verhaege K, Proc. 1994 EOS/ESD-Symp. p. 49, 1994

Very-Fast Transmission Line Pulsing of Integrated Structures and the Charged Device Model,
Gieser H and Haunschild M, Proc. 1996 EOS/ESD-Symp. p. 85, 1996

Advanced CMOS Protection Device Trigger Mechanisms During CDM,
Duvvury C and Amerasekera A, Proc. 1995 EOS/ESD-Symp. p. 162, 1995

Recommendations to Further Improvements of HBM ESD Component Level Test Specification,
Verhaege K et al, Proc. 1996 EOS/ESD-Symp. p. 40, 1996

14.4 Susceptibilty for ESD

Catastrophic and Latent Damages in GaAlAs Lasers Caused by Electrical Transients,
Sim S P et al, Journal of Appl. Phys. Vol 55, p. 3950, 1984

Sensitivity Study to EOS/ESD of Bipolar Integrated Circuits,
Urbieta A, Proc. 1986 EOS/ESD-Symp. p. 214, 1986

The Effect of ESD on LED III-V Materials,
Talbot J W, Proc. 1986 EOS/ESD-Symp. p. 238, 1986

Electrostatic Discharge Effects on Gallium Arsenide Integrated Circuits,
Rubelcawo A L et al, Proc. 1986 EOS/ESD-Symp. p. 159, 1986

Internal Chip ESD Phenomena Beyond the Protection Circuit,
Texas Instrument Techn. Journal, Vol 5, No 3, 1988

Electrostatic Discharge (ESD) Thresholds for GaAs Microwave Power FET Devices,
Andersen W and Chase E, Proc. 1988 EOS/ESD-Symp. p. 205, 1988

Electrostatic Discharge (ESD) Sensitivity of Thin-Film Hybride Passive Components,
Thiet the Lai, IEEE Trans. Components, Hybrides and Manuf. Techn., Vol 12, No 2, Dec 1989

Slow Transient Voltage and ESD Breakdown in Unprotected MOS Gate Oxides,
Tunnicliffe M J et al, Proc. European ESD Ass. Conf., 1991

Parametric Drift in Electrostatically Damaged MOS Transistors,
Tunnicliffe M J et al, Proc. 1992 EOS/ESD-Symp. p. 112, 1992

Theoretical Study of Expected EOS/ESD Sensitivity of III-V Compounds,
Chase E W, Proc. 1988 EOS/ESD-Symp. p. 65, 1988

Semiconductor Device Reliability,
Chrior A and Unger B A(Editors), Kluwer Academic Publishers, 1990

VLSI Interconnections Technology, Present and Future,
Saraswat K C, Proc. 1987 EOS/ESD-Symp. p. 1, 1987

14.5 Failure Analysis

Analysis of ESD Failures,
McAteer O J and Twist R E, Proc. 1981 EOS/ESD-Symp. p. 14, 1981

Failure Analysis Methodology for Revealing ESD Damages to IC's
Taylor R G et al, Proc. 1984 EOS/ESD-Symp. p. 189, 1984

Deficiencies in ESD Testing Methodology Highlighted by Failure Analysis,
Taylor R G et al, Proc. 1985 EOS/ESD-Symp. p. 141, 1985

Electro-Thermomigration as an Electrical Overstress Failure Mechanism,
Pierce D G, Proc. 1985 EOS/ESD-Symp. p. 67, 1985

Cost effective Failure Analysis Method for Detecting Failure Site Asscociated with Extremely Small Leakage,
Hansel G and Villareal E, Proc. 1987 EOS/ESD-Symp. P. 71, 1987

The Metallurgical Study of ESD Damage in 256 K DRAM Devices,
Jon M and Welsher T, Proc. 1987 EOS/ESD-Symp. p. 78, 1987

Electrostatic Damage in Electronics Devices and Systems,
Greason W D, John Wiley and Sons; N.Y., 1987

Non-Standard Failure Mechanisms Induced by Electrostatic Discharges,
Di Janni M et al, Proc. ESREF' 90 Conf. p 335, 1990

Characterization and Failure Analysis of Advanced CMOS Sub-Micron ESD Protection Structures,
Cook C and Daniel S, Proc. 1992 EOS/ESD-Symp. p. 149, 1992

A Statistical Method for the Detection of Gate Oxide Breakdowns Due to Fast EOS Events, such as ESD, on Power DIMOS Devices,
Bierman G, Proc. 1993 EOS/ESD-Symp. p. 143, 1993

Failure Analysis of Shallow Trench Isolated ESD Structures,
Never J and Voldman S, Proc. 1995 EOS/ESD-Symp. p. 273, 1995

ESD Failure Mechanisms of Inductive and Magnetoresistive Recording Heads,
Wallash A J and Hughbanks T S, Proc. 1995 EOS/ESD-Symp. p. 322, 1995

Acoustic Microscopy (A.M.) Application,
Carbone R A, Proc. ISTFA 92. p. 301, 1992

Characterization of Defects in Plastic Encapsulated Devices Using Acoustic Microscopy,
Semmens J E and Kessler L W, Proc. Meppe Focus, Jan. 1991, p 230, 1991

FIB, Focused Ion Beam Technology,
Sudraud P, Proc. 1991 European ESD Ass. Conf., Eindhoven, NL, 1991

Analysis of ESD Protection Networks for DMOS Power Transistors by means of Static and Time-resolved Emission-Microscopy,
Bonati B et al, Proc. ESREF' 92 p. 73, 1992

Field Components Reliability Analysis for French Telecommunications Equipment,
Boulaire J Y et al, Proc. ESREF' 92, p. 81, 1992

Latent Gate Oxide Damage Induced by Ultra-Fast Electrostatic Discharge (Thesis, ETH nr 11212, Zurich Switzerland),
Reiner J C, Series in Microelectronics, Vol 51, Hartung-Gorre Verlag, Konstanz, 1996

14.6 Latent Failures

Latency ESD Failures,
McAteer et al, Proc. 1982 EOS/ESD-Symp. p. 11, 1982

Latency Effects Observed in HCMOS Input Protection Circuits,
Crockett R G M et al, Proc. 1984 EOS/ESD-Symp. p. 196, 1984

Latent Failures Due to Electrostatic Discharge in CMOS Integrated Circuits,
Gammill P E and Soden J M, Proc. 1986 EOS/ESD-Symp. p. 75, 1986

Studies and Relevation of latent ESD-Failures,
Hellström S et al, Proc. 1986 EOS/ESD-Symp. p. 81, 1986

Latent ESD Metalization Damage,
Brennan T, Proc. 1987 ISTFA Conf. p. 177, 1987

Stress – Hardening of Damage Thresholds Due to Cumulative ESD Pulse Testing,
Hull R, Proc. 1988 EOS/ESD-Symp. p. 39, 1988

ESD Latency, A Failure Analysis Investigation,
Woodhouse J and Lowe K, Proc. 1988 EOS/ESD-Symp. p. 47, 1988

On Latency and Physical Mechanisms Underlying Gate Oxide Damage During ESD Events in N-Channel MOSFET's,
Krakauer D and Mistry K, Proc. 1989 EOS/ESD-Symp. p. 121, 1989

Investigation of Latent Failures Due to ESD in CMOS Integrated Circuits,
Greason W D and Chum K, Proc. 1991 EOS/ESD-Symp. p. 163, 1991

Study of Time-Dependent Latency in a Logical Microcircuit,
Hellström S and Welander A, Proc. ESREF' 91 p. 101, 1991

The Integrity of Gate Oxide Related to Latent Failures Under EOS/ESD conditions,
Greason W D and Chum K, Proc. 1992 EOS/ESD-Symp. p. 106, 1992

The Identification and Analysis of Latent ESD Damage on CMOS Input Gates,
Colvin J, Proc. 1993 EOS/ESD-Symp. p. 109, 1993

Latent Gate Oxide Damage Induced by Ultra-Fast Electrostatic Discharge (Thesis, ETH nr 11212, Zurich Switzerland),
Reiner J C, Series in Microelectronics, Vol 51, Hartung-Gorre Verlag, Konstanz, 1996

14.7 Indirect ESD

A Treatise on Electricity and Magnetism, Vol 1, 3rd Edition (1873).
Maxwell J C, 1873 (1974), Reprinted by Dover, New York, 1974

EMI Characteristics of ESD in a Small Air Gap,
Honda M and Kawamura T, Proc. 1984 EOS/ESD-Symp. p. 124, 1984

Electrostatic Spark Discharges: This Nuisance with 3 Faces,
Honda M and Ogura Y, Proc. 1985 EOS/ESD-Symp. p. 149, 1985

Standards and Regulations for Evaluating ESD Immunity at the System Level – An Update,
Dash G, Proc. 1988 EOS/ESD-Symp. p. 190, 1988

Evaluation of EMI Immunity Using Indirect ESD Testing,
Honda M, Proc. 1988 EOS/ESD-Symp. p. 185, 1988

Simulation of the EMP from ESD,
Boverie B, Proc. 1989 EOS/ESD-Symp. p. 145, 1989

Applying High Frequency Techniques to Measuring ESD Phenomena and Effects,
Smith D C, Proc. 1989 EOS/ESD-Symp. p. 114, 1989

ESD Stress on PCB Mounted IC's Caused by Charged Boards and Personnel,
Boxleitner W, Proc. 1990 EOS/ESD-Symp. p. 54, 1990

Modeling ESD Induced Potentials in Electronic Systems,
Greason W D, Proc. 1990 EOS/ESD-Symp. p. 74, 1990

Techniques and Methodologies for Making System Level ESD Response Measurements for Troubleshooting or Design Verification,
Smith D C, Proc. 1992 EOS/ESD-Symp. p. 47, 1992

Electrostatic Discharge Analysis for Spacecraft in Geosynchronous Orbit,
Perez R, Proc. 1993 EOS/ESD-Symp. p. 29, 1993

A New Type of Furniture ESD and its Implications,
Smith D C, Proc. 1993 EOS/ESD-Symp. p. 3, 1993

Clarification of Ultra-High-Speed Electrostatic Discharge and Unification of Discharge Model,
Tanaka M et al, Proc. 1994 EOS/ESD-Symp. p. 170, 1994

New Approaches to Indirect ESD Testing,
Honda M and Kinoshita T, Proc. 1995 EOS/ESD-Symp. p. 86, 1995

To What Extent Do Contact-Mode and Indirect ESD Test Methods Reproduce Reality?
Pommerenke D and Adam M, Proc. 1995 EOS/ESD-Symp. p. 101, 1995

What is a Real 1 Ghz Bandwidth ESD Generator Calibration?
Glatti P, Proc. 1994 EOS/ESD-Symp. p. 180, 1996

14.8 Other types of perturbations in electronics

High Frequency Measurements and Noise in Electronic Circuits,
Smith D C, Van Norstrand Reinhold, 1992

Powerline Disturbances and Electronic Systems – A Primer,
Harford J R, Proc. 1990 EOS/ESD-Symp., p. 19, 1990

Noise in Solid State Devices and Circuits,
Aldets van der Ziel et al, Wiley and Sons, N.Y., 1986

EURO-EMC' 90 CONFERENCE
Barbican, London, 1990

Quatum 1/F Noise and Other Low Frequency Fluctuations in Electronic Devices,
Handel P H and Chung A L, S:t Louis, Mo, AIP Conference Proceedings 282, 1992

Engineering Electromagnetic Compatibility: Principles, Measurements and Technologies,
Kodali V P, IEEE, Inc, N.Y., 1996

14.9 The impact of electric fields and electromagnetic waves on the organism

Many investigations were performed on this subject of which only a few are quoted here as examples.

Electrical Wire Configurations and Childhood Career,
Wertheimer N and Leeper E, Am. J. Epidemiol, 1979:109, p. 273, 1979

Invited Paper – Effects of Air Ions and electric Fields on Health and Productivity,
Carlton M D, Proc. 1984 EOS/ESD-Symp., p. 34, 1984

60 Hz Electric-Field Effects on Pineal Melantonin Rhytms: Time Course for Onset and Recovery,
Wilson B W et al, Bioelectromagnetics, 1986:7(2), p. 239, 1986

Work with Video Display Terminals Among Office Employees. I Subjective Symptoms and Discomfort,
Knave B G et al, Scand J. Works Environ. Health. 1985:11, p. 457, 1985

Work with Video Display Terminals Among Office Employees. II. Physical Exposure Factors,
Knave B G et al, Scand J. Work Environ. Health. 1985:11, p. 467, 1985

The Risk of Miscarriage and Birth Defects Among Women Who Use Visual Display Terminals During Pregnancy,
Goldhaber M K et al, Am J. Ind. Med. 1988:13, p. 695, 1988

Magnetic Field Exposure from Electric Appliances and Childhood Cancer,
Savitz D A et al, Am J. Epidemiol. 1990:131, p. 763, 1990

Magnetite in Human Tissues: a mechanism for the biological effects of weak ELF magnetic fields, *Kirschvink J L et al, Bio electromagnetics, Suppl. I: 101-13, 1992*

Biological Effects of Electromagnetic Fields and Radiation,
Saunders R D et al, J. Radiol. Prot. 1991:11, p. 27, 1991

Incidence of Leukaemia and Brain Tumours in Some "Electrical Operations",
Törnqvist S et al, Br J. Ind. Med. 1991:48, p. 597, 1991

Biological Effects of Extremely Low-Frequency Electromagnetic Fields: In Vivo Studies,
Anderson L E, Am. Ind. Hyg. Assoc. J. 1993:54, p. 186, 1993

Leukaemia and Brain Tumours in Norwegian Railroad Workers, a Nested Case – Control Study,
Tynes T et al, Am. J. Epidemiol, 1994:139, p. 645, 1994

Residential Proximity to Electricity Transmission and Distribution Equipment and Risk of Childhood Leukaemia, Childhood-Lymphona, and Childhood Nerves System Tumours: Systematic Review, Evaluation and Analysis,
Washburn E P et al, Cancer Causes and Control, 1994:5, p. 299, 1994

WHO Environmental Health Criteria 137: Electromagnetic Fields (300 Hz to 300 GHz),
World Health Organization, Geneva, 1993

14.10 Simulation methods

Determination of Threshold, Failure Levels of Semiconductor Diodes and Transistors Due to Pulse Voltages,
Wunsch D C and Bell R R, IEEE Trans. Nuclear Science Vol NS-15, p. 244, 1968

An Overview of EOS Effects on Semiconductor Devices,
Pierce D G and Durgin D L, Proc. 1981 EOS/ESD-Symp. p. 120, 1981

Semiconductor Junction Non-Linear Failure Power Thresholds: Wunch-Bell Revisited,
Ash M S, Proc. 1983 EOS/ESD-Symp., p. 122, 1983

Modeling and Testing for Second Breakdown Phenomena,
Orvis W J, Proc. 1983 EOS/ESD-Symp., p. 108, 1983

Transient Simulation of Silicon Devices and Circuits,
Bank R E, IEEE Trans. on Electron Devices, Vol ED-32, No 10, p. 1992, 1985

On Dielectric Breakdown in Oxidized Silicon,
Wolters D R and Zegers-van Duijnhoven, J. Vac. Sci. Technol, A 5(4), p. 1563, 1987

Numerical Simulation of Electrothermal Effects in ESD Protection Devices,
Hellström S et al, Proc. ESREF' 92, p. 77, 1992

Thermal Failure in Semiconductor Devices,
Dwyer et al, Solid State Electronics, Vol 33, No 5, p. 553, 1990

Simulation, a Tool for Designing in Reliability,
Brombacher A et al, Proc. ESREF' 92, p. 201, 1992

Physics of Electro-Thermal Effects in ESD Protection Devices,
Krabbenborg B et al, J. of Electrostatics, Vol 28, No 3, 1992

Three-Dimensional Transient Electrothermal Simulation of Electrostatic Discharge Protection Circuits,
Voldman S et al, Proc. 1994 EOS/ESD-Symp., p. 246, 1994

Simulation of Electrothermal Effects in Semiconductor Devices Exposed to ESD-Transients,
Hellström S, URSI General Assembly (XXIVth), Kyoto, Japan, p. 596, 1993

Circuit-Level Simulation of CDM-ESD and EOS in Submicron MOS Devices,
Ramaswamy S et al, Proc. 1996 EOS/ESD-Symp., p. 316, 1996

Understanding ESD Failure Threshold of N+/P Junetion by Electrothermal Simulation.
Buj C et al, Proc. European ESD Ass. Conf. 1993

Simulation of Transients in pn-junctions,
Hellström S and Eriksson L O, Proc. EUROSTAT 90, Barbican, London, 1990

Simulation of Thermal Runaway During ESD Events,
Beltman R A M et al, Proc. 1990 EOS/ESD-Symp., p. 157, 1990

Use of Electrothermal Simulation to Analyze Thermal Breakdown on $N^+/P/P^+$ *Diode During ESD Pulse,*
Buj C et al, Proc. ESSDERC' 93, 1993

14.11 Protection Methods

Electrostatic Measurement for Process Control,
Baumgartner G, Proc. 1984 EOS/ESD-Symp., p. 25, 1984

Triboelectric Charge its ESD Ability and a Measurement Method for its Propensity on Packaging Materials,
Huntsman J R, Proc. 1984 EOS/ESD-Symp., p. 64, 1984

Triboelectric Characterization of Packaging Materials,
Unger B A and Hart D L, Proc. EOS/ESD-Symp., p. 107, 1985

ESD Discharge Events from Handling Systems,
Bonnot G, European ESD Ass. Conf., 1991, Einhoven, NL.

Understanding Pink Poly,
Havens M R, Proc. 1989 EOS/ESD-Symp., p. 95, 1989

Contaminated Antistatic Poly Ethylene,
Denton R et al, Proc. 1987 EOS/ESD-Symp., p. 36, 1987

Static Survey Meters,
Lonborg J O, Proc. 1983 EOS/ESD-Symp., p. 63, 1983

Triboelectricity and Surface Resistivity do not Correlate,
Fowler S L, Proc. 1988 EOS/ESD-Symp., p. 103, 1988

Surface Resistivity and Static Decay do not Correlate,
Fowler S L, Proc. 1989 EOS/ESD-Symp., p. 7, 1989

Analysis of Electrostatic Charge Propensity of Floor Finishes,
Lingousky J F and Holt V E, Proc. 1983 EOS/ESD-Symp., p. 17, 1983

A Proposed Test Methodology for Evaluating the ESD Control Characteristics of Floor Materials,
Brandt M T and Halperin S, Proc. 1990 EOS/ESD-Symp., p. 253, 1990

Development of Highly Efficient, Static Dissipative Systems Based on Electroconductive Powders,
Lawton R S et al, Proc. 1991 EOS/ESD-Symp., p. 182, 1991

The Chemistry of Antistatic Additives,
Anderson B J, Proc. 1991 EOS/ESD-Symp., p. 199, 1991

A New Permanent ESD and Corrosion Resistant Material,
Franey J P et al, Proc. 1991 EOS/ESD-Symp., p. 210, 1991

Inherently Dissipative Polymer Films,
Fahey T E and Wilson G F, Proc. 1992 EOS/ESD-Symp., p. 189, 1992

Experimental Comparison of Methods of Charge Decay Measurements for a Variety of Materials,
Chubb J, Proc. 1992 EOS/ESD-Symp., p. 203, 1992

Testing ESD Shielding Bags with an Improved Bag Tester According to EIA-541,
Gaertner R and Schmeer H, Proc. 1992 EOS/ESD-Symp., p. 218, 1992

Triboelectric Charge-Discharge Damage Susceptibility of Large Scale IC's,
Lieske N and Swenson D, Proc. 1987 EOS/ESD-Symp., p. 274, 1987

A Method for Measurement of Triboelectric Charging,
Börjesson A, Proc. 1995 EOS/ESD-Symp., p. 253, 1995

The Physics of Charge Neutralization by Air Ions,
Jonassen N, Proc. 1986 EOS/ESD-Symp., p. 35, 1986

The Room Air Ionization System a Better Alternative than 40% Relative Humidity,
Mykkanen C F and Blinde D R, Proc. 1983 EOS/ESD-Symp., p. 67, 1983

Best practices for Applying Air Ionization,
Steinman A, Proc. 1995 EOS/ESD-Symp., p. 245, 1995

About the Effectiveness of Grounding Chains on Carts and the Unexpected Behaviour of Insulating Wheels,
Krog-Jensen E and Hjarn L, Proc. 1993 EOS/ESD-Symp., p. 165, 1993

A Design Methodology for ESD Protection Netword,
Pelella A R and Domingos H, Proc. 1985 EOS/ESD-Symp., p. 24, 1985

Design and Test Results for a Robust CMOS VLSI Input Protection Network,
Lin C M et al, Proc. 1986 EOS/ESD-Symp., p. 24, 1986

NMOS Protection Circuitry,
Rountree R N and Hutchins C L, IEEE Trans Electron Devices, ED-32:5, p. 910, 1985

Improving the ESD Failure Threshold of Silicided n-MOS Output Transistors by Ensuring Uniform Current Flow,
Polgreen T and Chatterjee A, Proc. 1989 EOS/ESD-Symp., p. 167, 1989

Input Protetion Design to Overall Chip Reliability,
Duvvury C et al, Proc. 1989 EOS/ESD-Symp., p. 190, 1989

Enhanced p^+ Substrate Tap Conductance in the Presence of NPN Snapback,
Maloney T J, Proc. 1990 EOS/ESD-Symp., p. 197, 1990

A New ESD Protection Concept for VLSI CMOS Circuits Avoiding Circuit Stress,
Guggenmos X and Holzner R, Proc. 1991 EOS/ESD-Symp., p. 74, 1991

MOS Integrated Circuits Measurement Method of Resistance to Electrostatic Discharges: Choice and Result (in French),
Levy P J, Proc. 4th Int. Conf. Reliability and Maintainability, Part 1, p. 65, 1984 (France)

Two Unusual HBM ESD Failure Mechanisms on a Mature CMOS Process,
Johnson C et al, Proc. 1993 EOS/ESD-Symp., p. 225, 1993

Advanced CMOS Protection Device Trigger Mechanism During CDM,
Duvvury C and Amerasekera A, Proc. 1995 EOS/ESD-Symp., p. 162, 1995

Designing MOS Inputs and Outputs to Avoid Oxide Failure in the Charged Device Model,
Maloncy T, Proc. 1988 EOS/ESD-Symp., p. 220, 1988

The Production Operator: Weak Link or Warrior in the ESD Battle?
Hansel G E, Proc. 1983 EOS/ESD-Symp., p. 12, 1983

A Systematic ESD Program Revisited,
Dangelmayer G T, Proc. 1992 EOS/ESD-Symp., p. 11, 1992

ESD Controls in Hazardous High Voltage Environments,
Berkowitz M B, Proc. 1990 EOS/ESD-Symp., p. 27, 1990

Avoiding Costly Pitfalls in Establishing an ESD Control Program,
Hensel W A, Proc. 1991 EOS/ESD-Symp., p. 41, 1991

You've Implemented An ESD program – What's next?
McFarland W and Brin R, Proc. 1993 EOS/ESD-Symp., p. 41, 1993

Implementing an ESD program in a Multinational Company: A Cross – Cultural Experience,
Tan W, Proc. 1994 EOS/ESD-Symp., p. 13, 1994

Effectiveness of ESD Training Using Multimedia,
Smalanskas G and Mason J, Proc. 1995 EOS/ESD-Symp., p. 129, 1995

ESD Control Program: A Viewpoint from the Receiving End,
Koay T and Tan W, Proc. 1995 EOS/ESD-Symp., p. 134, 1995

DOD-HDBK 263 A

DOD-STD-1686A, 1988

14.12 Failure frequency and costs for ESD

A Review of EOS/ESD Field Failures in Military Equipment,
Green T and Denson W, Proc. 1988 EOS/ESD-Symp., p. 7, 1988

ESD Control Implementation and Cost Avoidance Analysis,
Downing M H, Proc. 1983 EOS/ESD-Symp., p. 6, 1983

Plastic Encapsulated IC's in Military Equipments Reliability Prediction Modeling,
Brizoux M et al, Proc. ESREF' 91, p. 847, 1991

Reliability Behaviour of Electronic Components as a Function of Time,
Campbell et al, Proc. ESREF' 91, p. 41, 1991

ESD Protection Measures Return on Investment Calculation and Case Study,
Helling K, Proc. 1996 EOS/ESD-Symp., p. 130, 1996

Extent and Cost of EOS/ESD Damage in an IC Manufacturing Process,
Wagner R et al, Proc. 1993 EOS/ESD-Symp., p. 49, 1993

Reducing Field Failure Rate with Improved EOS/ESD Design,
Euzent B L et al, Proc. 1991 EOS/ESD-Symp., p. 59, 1991

What Happens After it Leaves the Plant?
Rataski C, Proc. 1991 EOS/ESD-Symp., p. 65, 1991

Tracking Results of an ESD Control Program within a Telecommunications Service Company,
Zezulka R J, Proc. 1989 EOS/ESD-Symp., p. 36, 1989

An ESD Expert System,
Fuqua N and Dekkers D, Proc. 1987 EOS/ESD-Symp., p. 129, 1987

Tape and Reel Packaging - An ESD Concern,
Jon M C et al, Proc. 1988 EOS/ESD-Symp., p15, 1988

NARTE: The National Association of Radio and Telecommunications Engineers, Inc., has a certification program for ESD, EMC and Telecommunication Technicians and Engineers.

Some relevant books:

Quality and Reliability of Technical Systems, Theory - Practice - Management,
Birolini A, Springer-Verlag, 1994

Field and Wave Electromagnetics Second Edition,
Cheng D K C, Addison-Wesley Publishing Company, 1989

ESD Program Management,
Dangelmayer G T, Van Nostrand Reinhold, N.Y:, 1990

Reliability and Degradation: Semiconductor Devices and Circuits,
Howes M J and Morgan D V, Editors, John Wiley and Sons, 1981

Electronic Component Reliability,
Jensen F, Wiley, 1995

Electromagnetic Compatibility and Telecommunications,
Lindstrom E R and Horlich J, 1995, NIST Handbook 105:11, US Dept. of Commerce, 1995

Electrostatic Discharge Control,
McAteer O J, McGraw Hill, 1990

Electrostatic Discharge in Electronics,
Greason W D, John Wiley and Sons, 1992

Total Quality Management in the Electronics Industry: A User's and Supplier's Guide,
Beasley K, Technical communications (Publishing) Ltd, 1992

The US ESD-Ass. Standards can be obtained from:
ESD Association
7900 Turin Rd., Bldg. 3, Stc 2,
Rome, N Y 13440-2069
USA

Springer and the environment

At Springer we firmly believe that an international science publisher has a special obligation to the environment, and our corporate policies consistently reflect this conviction.

We also expect our business partners – paper mills, printers, packaging manufacturers, etc. – to commit themselves to using materials and production processes that do not harm the environment. The paper in this book is made from low- or no-chlorine pulp and is acid free, in conformance with international standards for paper permanency.

Printing: Mercedesdruck, Berlin
Binding: Buchbinderei Lüderitz & Bauer, Berlin